A PRIMER ON DETERMINISM

A PALLAS PAPERBACK

JOHN EARMAN

Dept. of History and Philosophy of Science
University of Pittsburgh

A PRIMER
ON DETERMINISM

D. REIDEL PUBLISHING COMPANY

A MEMBER OF THE KLUWER ACADEMIC PUBLISHERS GROUP

DORDRECHT / BOSTON / LANCASTER / TOKYO

Library of Congress Cataloging-in-Publication Data

Earman, John.
    A primer on determinism.

    (University of Western Ontario series in philosophy of
science ; v. 32)
    Bibliography: p.
    Includes index.
    1. Determinism   (Philosophy) 2.   Physics—Phi-
losophy.   3. Science—Philosophy.   I. Title.   II.
Series.
QC6.4.D46E27 1986   123   86–15609
ISBN 90–277–2240–4
ISBN 90–277–2241–2 (pbk.)

Published by D. Reidel Publishing Company
P.O. Box 17, 3300 AA Dordrecht, Holland

Sold and distributed in the U.S.A. and Canada
by Kluwer Academic Publishers,
101 Philip Drive, Assinippi Park, Norwell, MA 02061, U.S.A.

In all other countries, sold and distributed
by Kluwer Academic Publishers Group,
P.O. Box 322, 3300 AH Dordrecht, Holland

*Also published in 1986 in hardbound edition by Reidel
in the Western Ontario Series in the Philosophy of Science*

Printed in The Netherlands

For Marvel and Judy, two ladies of the night,
And Hisser and Larry, two gentlemen of different stripes.

TABLE OF CONTENTS

PREFACE

The title of this work is to be taken seriously: it is a small book for teaching students to read the language of determinism. Some prior knowledge of college-level mathematics and physics is presupposed, but otherwise the book is suitable for use in an advanced undergraduate or beginning graduate course in the philosophy of science. While writing I had in mind primarily a philosophical audience, but I hope that students and colleagues from the sciences will also find the treatment of scientific issues of interest.

Though modest in not trying to reach beyond an introductory level of analysis, the work is decidedly immodest in trying to change a number of misimpressions that pervade the philosophical literature. For example, when told that classical physics is not the place to look for clean and unproblematic examples of determinism, most philosophers react with a mixture of disbelief and incomprehension. The misconceptions on which that reaction is based can and must be changed. In addition to embodying falsehoods about the implications of modern physics for the truth of the doctrine of determinism, the literature also fails to convey a sense of how determinism actually works or fails to work in physics; how delicate the doctrine is in some respects, requiring a host of enabling assumptions for it to stand a chance of being true, and yet how robust it is in other respects, not only surviving but absorbing and growing stronger from potential counterexamples; how straightforward and mundane it is in some ways, and yet how subtly and exquisitely connected it is to a complex of issues about the nature of space and time, about the ultimate furniture of the universe, and about the concept of physical possibility. Further, the meaning of determinism is often mangled in the process of philosophical ax-grinding, the worst offenders being those Libertarians who, fearful that determinism will undermine human freedom and dignity, are careful to define it so that it can't be true. And finally, and most distressingly of all, the literature conveys the false impression that outside of the quantum domain determinism is a dead issue, whereas in fact it continues to throw off new problems and challenges for philosophy and for science. The main burdens of this Primer, then, are to say what determinism is and to describe in a way that is both introductory and realistic the problems and prospects of determinism in various branches of modern physics and at the same time to reveal how advances in understanding those problems and prospects are bound up with advances in mathematics, physics, philosophy of science, and metaphysics. I do not claim to have succeeded but only to have made a long overdue start on this neglected task. The challenge of completing the task will have to be left to abler hands.

It is not possible for me to name all of the people who have helped me in my struggle to come to grips with determinism. But I would be remiss if I did not give special thanks to C. A. Anderson, A. Fine, R. Geroch, C. Glymour, M. Hamermesh, P. Kitcher, D. Malament, and L. Markus. I am grateful to D. Reidel and R. Bogdan, editor of *D. M. Armstrong* (Dordrecht: D. Reidel, 1984), for permission to use material which appeared in my article "Laws of Nature: The Empiricist Challenge." Finally, I am proud to say that I was not supported by a Grant or any form of largesse, public or private.

November, 1985   J. E.

CHAPTER I

INTRODUCTION

> With earth's first clay they did the last man knead,
> And there of the last harvest sowed the seed.
> And the first morning of creation wrote
> What the last dawn of reckoning shall read.

Determinism is a perennial topic of philosophical discussion. Very little acquaintance with the philosophical literature is needed to reveal the Tower of Babel character of the discussion: some take the message of determinism to clear and straightforward while others find it hopelessly vague and obscure; some take determinism to be intimately tied to predictability while others profess to see no such bond; some take determinism to embody an *a priori* truth, others take it to express a falsehood, and still others take it to be lacking in truth value; some take determinism to undermine human freedom and dignity, others see no conflict, and yet others think that determinism is necessary for free will; and on and on. Here we have, the cynic will say, a philosophical topic *par excellence*!

Without any touch of cynicism one may ask what yet another tour of this Babel can hope to accomplish, save possibly to add another story to the Tower. My answer is not at all coy. Essential to an understanding of determinism is an appreciation of how determinism works or fails to work in physics, the most basic of all the empirical sciences; but it is just this appreciation I find lacking in the philosophical literature. My complaint does not center on technical niceties, though they become crucial at various junctures, but on the larger picture which emerges from the philosophical and the popular science literature. Classical physics is supposed by philosophers to be a largely deterministic affair and to provide the paradigm examples of how determinism works. Relativity theory, in either its special or general form, is thought merely to update classical determinism by providing for Newtonian mechanisms relativistic counterparts that are no less and no more deterministic. And it is only with the advent of the quantum theory that a serious challenge to determinism is supposed to emerge; the challenge is not simply that quantum mechanics is *prima facie* non-deterministic

1

but that "no hidden variable" theorems show that, under plausible constraints, no deterministic completion of the quantum theory is possible.

This picture is badly out of focus. Newtonian physics, I will argue, is not a paradise for determinism; in fact, Newtonian worlds provide environments that are quite hostile to determinism, and some of the alleged paradigm examples of Newtonian determinism are not examples of determinism at all, at least not without the help of props which sometimes have a suspiciously artificial and even question-begging character. The special theory of relativity rescues determinism from the main threat it faces in Newtonian worlds, and in special relativistic worlds pure and clean examples of determinism, free of artificial props, can be constructed. However, the general theory of relativity poses new and even graver challenges, challenges which are currently being addressed on the frontiers of scientific research. The quantum theory, of course, poses challenges of its own; but the first and foremost challenge is not to the truth of the doctrine of determinism but to its meaning in quantum worlds where the ontology may be nothing like that presupposed in the Newtonian and relativistic formulations of the doctrine.

Before we can have a Treatise on determinism, we need a Primer; we need first to learn how to spell and read the language of determinism before attempting lofty dissertations. The plan of this primer is as follows. In Ch. II I briefly review some of the many philosophical attempts to define determinism, and I propose a conception which, though vague, is nonetheless useful as a starting point for the discussion of classical and relativistic physics. Ch. III surveys the trials and tribulations that determinism faces in Newtonian worlds. Ch. IV is devoted to special relativistic physics and shows how this context makes possible the fulfillment of the classical dream of determinism, a dream which is disturbed by nightmares in Newtonian worlds. Ch. V interrupts the discussion of specific developments in physics to get a better fix on the crucial but troublesome notion of laws of nature. Ch. VI discusses the connections among determinism, mechanism, and effective computability. Computability is initially taken in Turing's sense and various results about the preservation of computability of initial data under deterministic evolution are reviewed; but it is also urged that deterministic systems give rise to a much broader notion of effective computability of which Turing's is but a special case. Ch. VII charts the relation of determinism to various time symmetries of laws, especially time reversal invariance, periodicity, and time translation invariance. Ch. VIII offers some suggestions for understanding the concepts of randomness and chaos and the conditions under which random and chaotic behavior are and are not able to peacefully coexist with determinism. Ch. IX introduces the notion of instability, or sensitive dependence on initial conditions, and shows how it forms one of the bridges leading from determinism on the micro-level to apparent randomness on the macro-level. Chs. X and XI complete the survey of the implications of modern physics for determinism by discussing in turn the general theory of relativity and the quantum theory. Finally, Ch. XII offers a few comments on the determinism-free will controversy. My main aims here are to show what is at stake and to diagnose the reasons why the issues have seemed so intractable.

As befits a primer, not much is demanded of the reader in terms of scientific knowledge, for while a large number of illustrative examples are drawn from mathematical physics, their presentation is designed to make clear the import with only the minimum of technical detail. At the same time it has to be recognized that determinism has been fruitful in suggesting mathematical problems, and the success or failure of determinism can in turn depend on the delicate details of these problems. I try to indicate where this is so and refer the reader who wants to learn more to the relevant scientific literature. The real demand is on the reader's tolerance for gear shifting, for the presentation often swings from traditional philosophy of science to mathematics to speculative metaphysics and back to practical physics. These are not strands which can be woven neatly together, but as I will try to show, all of them must be grasped simultaneously if we are to get a real feel for determinism.

Whether or not the reader agrees with the specific conclusions and morals I draw, I hope that he will come away with an appreciation of how exquisitely subtle a doctrine determinism is; how robust it is in some ways and how fragile it is in others; and, most of all, how exciting and how alive it is, despite the attempts of philosophers to relegate it to museum status.

SUGGESTED READINGS

Wesley Salmon's (1971) "Determinism and Indeterminism in Modern Science" provides a quick and very readable survey of the received philosophical opinion on this topic. Some useful historical background can be found in Richard Taylor's (1967) article "Determinism" in the *Encyclopedia of Philosophy*.

CHAPTER II

DEFINING DETERMINISM

> Clearly our first problem must be to define the issue,
> since nothing is more prolific of fruitless controversy
> than an ambiguous question.
>
> (Bertrand Russell, "Determinism and Physics")

Russell's advice seems the essence of good sense. But in trying to heed it we find ourselves in a Catch-22 situation: we cannot begin to discuss the implications of physics for the truth of the doctrine of determinism until we know what determinism is; on the other hand, no precise definition can be fashioned without making substantive assumptions about the nature of physical reality, but as we move from classical to relativistic to quantum physics these assumptions vary and the definition of determinism must, to some degree, covary with them.

If we cannot begin with a definition that is at once precise and general, then either precision or generality must be dropped. My suggestion is the seemingly perverse one that initially we drop both. The starting definition I will recommend is vague — as befits a vague doctrine — and is aimed towards classical physics — as befits the historical origins of the doctrine. The advantage of this approach is that it provides a common thread linking disparate material; all of the detailed, technical conceptions to be discussed in succeeding chapters can be seen as attempts to make precise the basic intuitive idea or else to modify it so as to fit some new development in physics.

1. CLASSICAL DETERMINISM: THE VISION AND THE CONTEXT

Before turning to various attempts to define determinism, it is important to have before us the vision which these definitions seek to capture. A most vivid rendering of the vision was given by William James in an 1884 lecture to the Harvard Divinity School:

What does determinism profess? It professes that those parts of the universe already laid down absolutely appoint and decree what the other parts shall be. The future has no ambiguous possibilites hidden in its womb: the part we call the present is compatible with only one totality. Any other future complement than the one fixed from eternity is impossible. The whole is in each and every part, and welds it with the rest into an absolute unity, an iron block, in which there can be no equivocation or shadow of turning. (1956, p. 150)

The context of this vision is what I will call the classical world picture. Exactly what this means will be discussed in detail in the next chapter, but for now suffice it to say that the spatio-temporal structure of the world is assumed to embody an absolute or observer-independent simultaneity; 'the world-at-a-given-time' is, therefore, an invariantly meaningful concept. Further, at each instant, the state of the world is fully characterized by specifying the values of relevant physical magnitudes — instantaneous values of the positions and velocities of particles, instantaneous values of electric and magnetic field vectors, and the like. The context of this vision is thus broad enough to encompass both particles and fields, materialistic and non-materialistic ontologies. Determinism as it is understood here does not assume materialism or mechanism in any narrow sense; indeed, the magnitudes to be considered may be ones traditionally taken as 'mentalistic', and all that is required is that they be physicalistic in the minimal sense that they have a spatio-temporal representation.

2. WHAT DETERMINISM IS NOT: CAUSE AND EFFECT

Just as it is hard to make bricks without straw, so it is hard to do philosophy without straw men. Unfortunately, there are no out and out straw-man definitions of determinism. But there are some venerable definitions which are worthy of some philosophical bayonet practice and which also have the more positive virtue of pointing the way towards more adequate definitions.

Perhaps the most venerable of all the philosophical definitions holds that the world is deterministic just in case every event has a cause. The most immediate objection to this approach is that it seeks to explain a vague concept — determinism — in terms of a truly obscure one — causation. If we can achieve an analysis of determinism without explicit appeal to the notion of cause and effect, then that analysis is to be preferred to the one in question. A related objection concerns the lack of a perspicuous connection between the causation definition and James' sense of determinism. In one direction the connection can be made tight: if the world displays Jamesian determinism then the "Every event has a cause" can be vouchsafed by taking (as Laplace suggested —

see below) the state of the world at any moment as the cause of the state to follow. But in the other direction the connection is obscure: How does it follow as a result of every event's having a cause that the future has no ambiguous possibilities hidden in its womb? Perhaps the cause-effect relation can be explained in such a way that this implication becomes transparent; but it is that explanation we want and not the evocative but obscure formula "Every event has a cause." There is a reasonably precise explanation of cause-effect in terms of a causal chain or signal, i.e., the propagation of a disturbance, say, in the form of the continuous transmission of a quantity of mass or energy through space; but this explanation does not yield the desired result. Imagine a materialistic world consisting of massive particles whose trajectories are straight lines except where the trajectories happen to intersect. Every interesting event or happening in this world is a happening to a particle, viz., a change of position, a collision, etc. And every such event has a cause in terms of the earlier events on the causal chain or chains on which it lies. Yet this world may or may not be deterministic in James' sense; for it seems consistent with the description I have given that many future complements are compatible with the present state of this world.[1]

The proponent of the "Every event has a cause" formula may complain that I have not used an appropriate sense of 'cause', an appropriate one being one on which the formula has the force of "Every event has a cause *and* same causes always produce the same effects." Or alternatively, it might be conceded that the second half of the expanded formula does not follow from the meaning of 'cause' and simply has to be postulated. In either case I would agree that the expanded formula comes much closer to supplying a sufficient condition for determinism. But I also believe that the valid kernel of the expanded formula can be retained while stripping off the chaff of 'cause', 'effect', and 'produce'.

In philosophical parlance 'causality' is an ambiguous term, referring both to determinism and to the cause-effect relation. I suggest that this term either be shelved or else that it be reserved for determinism while 'causation' is used to name whatever goes on when one event causes another. The remainder of this book is devoted to a discussion of causality with only a few hesitant and apologetic references to causation.

3. PREDICTABILITY: LAPLACE'S DEMON

Pierre Simon Laplace offered a definition of determinism which starts with a causal flavor but ends by equating determinism with predictability.

We ought to regard the present state of the universe as the effect of its antecedent state and as the cause of the state that is to follow. An intelligence knowing all the forces acting in nature at a given instant, as well as the momentary positions of all things in the universe, would be able to comprehend in one single formula the motions of the largest bodies as well as the lightest atoms in the world, provided that its intellect were sufficiently powerful to subject all data to analysis; to it nothing would be uncertain, the future as well as the past would be present to its eyes. The perfection that the human mind has been able to give to astronomy affords but a feeble outline of such an intelligence. Discoveries in mechanics and geometry, coupled with those in universal gravitation, have brought the mind within reach of comprehending in the same analytical formula the past and the future state of the system of the world. All of the mind's efforts in the search for truth tend to approximate the intelligence we have just imagined, although it will forever remain infinitely remote from such an intelligence.[2]

While Laplace's approach comes closer to the mark than does the previous one, the appeals to an 'intelligence' (or 'demon' as it is often called) and to the concept of knowledge ought to sound warning bells. Depending upon what powers we endow the demon with, we get different senses of determinism. Endow it with the powers of the latest Cray computer or even with the powers of a universal Turing machine and we get a fairly interesting sense of determinism; but we also get a sense in which it is fairly certain that the universe is 'non-deterministic' in that future states are not always computable from present states, and this may be so even if the universe fulfills James' vision (see Ch. VI). Endow the demon with God-like powers and this difficulty is overcome, but only at the expense of the opposite difficulty; for now the demon will be able to foresee the future — to it no future event will be uncertain — but this foresight may be a reflection of its precognitive abilities rather than any deterministic feature of the world.

It could be replied that the intent of Laplace's definition is in the right direction and all that needs to be done is to cleanse it of any reference to a predictor. I applaud this sentiment, but I would go even further in recommending that the notion of prediction with all of its epistemological connotations be dropped altogether. The history of philosophy is littered with examples where ontology and epistemology have been stirred together into a confused and confusing brew. The Jamesian vision we are seeking to capture is an ontological vision; whether it is fulfilled or not depends only on the structure of the world, independently of what we do or could know of it. Of course, ontological determinism does have epistemological implications and these will be discussed in the appropriate places. But let us not confuse the implications of the doctrine with the doctrine itself. And let us resist the temptation to manufacture 'senses' of determinism. Producing an 'epistemological sense' of determinism is an abuse of language since we already have a perfectly adequate and more accurate term — prediction — and it also invites potentially misleading argumentation — e.g., in such-and-such a case prediction is not possible and, therefore, determinism fails. The most notorious form of this argument is due to Sir Karl Popper.

4. PREDICTABILITY: POPPER'S DEMON

'Scientific determinism' in Popper's sense is

the doctrine that the state of any closed physical system at any given future instant of time can be predicted, even from within the system, with any specified degree of precision, by deducing the prediction from theories, in conjunction with initial conditions whose required degree of precision can always be calculated (in accordance with the principle of accountability) if the prediction task is given. (1982, p. 36)

Popper's basic demand is that Laplace's demon should be construed "not as an omniscient God, merely as a super-scientist". This means that

The demon, like a human scientist, must *not* be assumed to *ascertain initial conditions with absolute mathematical precision*; like a human scientist, he will have to be content with a finite degree of precision. (1982, p. 34)

Also, the demon must be able to predict from within the system; that is, the demon is not construed as a disembodied spirit but, like a human scientist, must be assumed to belong to and to interact with the system whose future it is trying to predict. The 'principle of accountability' is imposed to assure that the required degree of precision for the initial conditions can be known beforehand so that a prediction which fails to meet the specified error limits cannot be dismissed on the grounds that the initial data were not accurate enough.

Actually, the first part of Popper's demand is sufficient by itself to allow Popper to reach the conclusion that, contrary to widespread belief, classical physics exhibits systems which are not deterministic. I myself reach a similar conclusion, but for quite different sorts of cases and for quite different reasons. This matter will be taken up in some detail in the following chapter. The point I wish to emphasize here is that the examples Popper uses to illustrate his conclusion serve only as a *reductio* of his definition of determinism. The combination of a strong form of instability, where small changes in initial conditions can give rise to large changes in future states, and the inability of the demon to ascertain initial conditions with mathematically exact precision can lead to a breakdown in prediction. But the proper conclusion to be drawn from this result is not that determinism fails but rather that determinism and prediction need not work in tandem; for the evolution of the system may be such that some future states are not predictable (at least not under Popper's strictures) although any future complement than the one fixed from eternity is impossible. Hadamard, the authority whom Popper cites on these matters, puts the point this way: if the future state does not depend continuously on the initial state, then "Everything takes place, physically speaking, as if the knowledge of . . . [the initial] data would *not* determine the unknown function" (1952, p. 38). Popper's definition cancels the crucial "as if".

Popper's view is particularly awkward in the case of classical statistical mechanics because it has the effect of brushing aside one of the central foundations problems; namely, how can the 'random' and 'chaotic' behavior exhibited on the macro-level by, say, a box of gas be reconciled with the micro-determinism of the gas molecules? After many decades of research, it has become apparent that a large part of the answer lies precisely in instability (see Ch. IX).

Why is such an acute philosopher as Sir Karl bent on using such a wrong-headed conception of determinism? (For those unfamiliar with the history of philosophy, I note that analogous questions arise for every Great Man.) Popper's avowed purpose in *The Open Universe* is to "make room within physical theory . . . for indeterminism" (1982, p. xxi). By construing determinism in terms of finite prediction tasks, Popper is able to achieve his goal, but the form of indeterminism he generates does not resolve the 'nightmare of physical determinism' of which he spoke so eloquently in "Of Clouds and Clocks." If physical determinism holds and the antecedent state of the universe suffices to fix the future physical state, including all of our movements and thus all of our actions, then "all our thoughts, feelings, and efforts can have no practical influence upon what happens in the physical world: they are, if not mere illusions, at best superfluous by-products ('epiphenomena') of physical events" (1972, p. 217); the whole world with everything in it would be a huge automaton and we would be "nothing but little cog-wheels, or at best sub-automata, within it" (p. 222). But if that is the nightmare, it would seem to persist even after it has been shown that the deterministic unfolding of physical events cannot be exactly charted by a Popperian demon because, for example, the unfolding is unstable.

Nor can Popper's definition of 'scientific determinism' be justified by combining the desire to make determinism a scientific doctrine with Popper's thesis that falsifiability provides the demarcation between science and nonscience. Also needed is the notion that to be falsifiable determinism must be construed as an assertion about finite prediction tasks, a notion which is contrary to the spirit of Popper's original liberal interpretation of the falsifiability criterion.

Unlike Popper and his kindred spirits in logical empiricism, I am not afraid to attach the label 'metaphysical' to the doctrine of determinism; indeed, it seems to me that determinism as James, Laplace, and others understand it is *both* 'scientific' *and* 'metaphysical'. Like Popper, I am interested in how scientific evidence, reasoning, and inference can be brought to bear on the doctrine of determinism. But the ties that bind determinism to hard empirical evidence (however that is taken) are far too complex, subtle and tenuous to be encapsulated in a tidy formula couched in terms of falsifiability, verfiability, testability, or the like. Nevertheless, I hold that the evidential grounding of determinism is not mysterious, or at least it is no more mysterious than the grounding of many other high level scientific claims, viz., that total energy is conserved or that the temporal evolution of the world is time reversible. And the 'scientific' status of these claims surely does not turn on construing them as claims about prediction tasks that can be carried out by embodied super-scientists interacting with the systems whose futures they are trying to predict.

5. RUSSELL'S DEFINITION

Russell's essay "On the Notion of Cause" can usefully be viewed as an attempt to carry out the recommended cleansing of Laplace's definition of its epistemological components so as to produce a purely ontological formulation. Here is the upshot of Russell's housecleaning.

A system is said to be 'deterministic' when, giving certain data, $e_1, e_2, \ldots, e_n$ at times t_1, $t_2, \ldots, t_n$ respectively, concerning this system, if E_t is the state of the system at any time t, there is a functional relation of the form

$$E_t = f(e_1, t_1, e_2, t_2, \ldots, e_n, t_n).$$

The system will be 'deterministic throughout the given period' if t, in the above formula, may be any time within that period . . . . If the universe, as a whole, is such a system, determinism is true of the universe; if not, not. (1953, p. 398)

This seems cogent enough at first reading, but the definition has, as Russell goes on to show, a very counterintuitive upshot. To illustrate, imagine a very simple universe containing a single dimensionless particle, and suppose that the state of the particle at any instant t is specified by its position coordinates x_t, y_t, z_t. The motion of the particle through space can be as complicated as you like as long as it can occupy only one place at a time. Then, as a matter of mathematical fact, there must exist functions f_1, f_2, f_3 such that $x_t = f_1(t)$, $y_t = f_2(t)$, $z_t = f_3(t)$. The example can be made more realistic by adding other particles and additional state variables, but the essential point remains the same. In Russell's own words:

It follows that, theoretically, the whole state of the material universe at time t must be capable of being exhibited as a function of t. Hence our universe will be deterministic in the sense defined above. But if this be true, no information is conveyed about the universe in stating that it is deterministic. (1953, p. 401)

Combining Russell and Popper, we have the first intimation of the Scylla and Charybdis between which determinism is forced to sail: tack one way in defining determinism and determinism wrecks on obvious falsity; tack the other way and it wrecks on triviality. Much of the later chapters will be devoted to the question of whether a clear course can be charted between these obstacles. The question is particularly thorny because it is not merely a matter of reading the answer off the relevant parts of physics, for the interpretation of the physics may turn in part on convictions about the form the answer should take.

Russell considers two suggestions for avoiding the trivialization of determinism. The first is to require that the Russell function be simple. This suggestion is quickly discarded; and rightly so since the connection between simplicity and determinism is indirect and tenuous. The world can be as simple as you like in its contents and temporal evolution and yet non-deterministic in James' sense; or it can be highly complex but leave no room for equivocation or shadow of turning in its future development.[3] The second suggestion is that time not be allowed to enter explicitly into the Russell function. Russell is able to point to an independent motivation for this restriction; namely, the belief in the "uniformity of nature," meaning that "no scientific law involves time as an argument, unless, of course, it is given in integrated form, in which case *lapse* of time, though not absolute time, may appear in the formulae" (1953, p. 401). But one can doubt whether uniformity of nature in Russell's sense is any more essential to determinism than is simplicity. Imagine a world in which the gravitational 'constant' is not constant but varies with time. Does such a time dependence automatically make the world nondeterministic, open to ambiguous future possibilities? This and other questions about the relation between determinism and time symmetries deserve careful scrutiny; some will be provided in Ch. VII, but for present purposes we can avoid the issue. For even if we grant Russell his "uniformity of nature" as regards laws of motion, it hardly follows that the Russell function will not involve time explicitly; indeed, if position changes with time, the Russell function can hardly avoid having time as an argument. Russell has confused a property of laws with a property of Russell functions.

6. WHAT DETERMINISM IS

At several points Russell refers, as we have just done, to laws of nature, and in the statement of the trivialization result he concludes from the existence of the Russell function that "the material universe *must* be subject to laws" (1953, p. 401). But on the usual understanding of natural laws, this is a non-sequitur. Laws may prohibit some instantaneous states, but the familiar dynamical laws of physics, or what have passed for them, typically allow a wide range of instantaneous states for any given system; viz., any set of non-coincident positions and (finite) velocities is an allowable state in Newtonian particle mechanics. The main bite of the dynamical laws comes in restrictions on the temporal transition from one allowable state to another. Laplacian determinism is a very special and very strong form of such a restriction: for any time t_1 and t_2 and any allowed state at t_1, there is one and only one allowed state at t_2.

This idea could be reexpressed in terms of the existence of a new type of Russell function F, now construed as a map from triples of allowed instantaneous states $\times$ time $\times$ time to allowed states. Read $s' = F(s, t_1, t_2)$ as: s' is the (unique) allowed state at t_2 when the state at t_1 is s, where s ranges over all allowed instantaneous states.[4] But the only gain in such a formulation is in pedanticism.

Instead of using Russell functions I prefer a more pictorially appealing approach based on the now fashionable notion of possible worlds. A 'world' here means a four-dimensional space-time world, the actual world being the collection of all events that have ever happened, are now happening, or ever will happen, and a possible world being a collection of possible events representing possible alternative histories to that of the actual world. The starting assumption is that these events can be fitted into the classical world picture: their spatio-temporal relations conform to the structure required in classical physics and the events themselves can be analyzed, for example, as changes in spatio-temporal magnitudes. Taking space-time rather than instantaneous states as the basis of analysis becomes unavoidable in relativistic physics, but it is equally useful in the discussion of determinism in Newtonian physics, as I will try to show in the following chapter.

Letting $\mathcal{W}$ stand for the collection of all physically possible worlds, that is, possible worlds which satisfy the natural laws obtaining in the actual world, we can define the Laplacian variety of determinism as follows. The world $W \in \mathcal{W}$ is *Laplacian deterministic* just in case for any $W' \in \mathcal{W}$, if W and W' agree at any time, then they agree for all times. By assumption, the world-at-a-given-time is an invariantly meaningful notion and agreement of worlds at a time means agreement at that time on all relevant physical properties. This concept of determinism can be broken down into two subconcepts. A world $W \in \mathcal{W}$ is *futuristically* (respectively, *historically*) *Laplacian deterministic* just in case for any $W' \in \mathcal{W}$, if W and W' agree at any time then they agree for all later (respectively, earlier) times.

Determinism needn't be an all-or-nothing affair. A world may be *partially deterministic*, deterministic with respect to some magnitudes (agreement on the values of which at any time forces agreement at other times) but not with respect to others. But while such a bifurcation is imaginable, it can produce tensions. Try, for example, to imagine that the world is only partially deterministic because it is deterministic only with respect to the magnitudes which characterize the ordinary matter of which we and our scientific instruments are composed but not with respect to the magnitudes which characterize the behavior of a free-spirited species of particle, the freeon (say). But either the freeon magnitudes interact with ordinary magnitudes or not. In the latter case the freeons are scientifically suspect entities since as far as science can teach us they are unknowable ghosts in the deterministic machine. In the former case it is hard to see how, without a cosmic conspiracy, the partial determinism for the ordinary magnitudes can be maintained since otherwise the non-deterministic evolution of the freeons would infect the evolution of ordinary matter. A concrete example of this tension will be examined in Ch. IV.

The world might be non-deterministic but still *conditionally deterministic* on a subset of magnitudes: if two worlds agree for all times on the values of the conditioning magnitudes and if they agree at any instant on the values of the other magnitudes, then they agree at any other instant. Faith in strict determinism and the discovery of conditional determinism will prompt the search for additional laws that determine the evolution of the conditioning magnitudes, thereby removing the condition and restoring determinism *simpliciter.*

It might be charged that the possible worlds analysis is a fraud: it is no more than a transcription of James' poetic vision into terms which are devoid of James' eloquence but which display not much compensating gain in clarity and precision. I couldn't agree more! But I also think that without prejudging detailed substantive issues in physics we cannot do much better for a direct ontological formulation of what is, after all, an ontological doctrine. The usefulness of the possible worlds formulation as a starting point for the discussion of these issues and their bearing on determinism will, I hope, become apparent in succeeding chapters. However, honesty also demands a confession of some of the potential pitfalls of the approach.

7. FEAR AND LOATHING

A principal reason for rejecting Russell's approach was the fear that, without the aid of artificial props, it would reduce determinism to a triviality. A similar fate awaits the possible worlds definition of determinism unless the properties which characterize the instantaneous state of the world are suitably restricted; in particular, they must be non-indexical and genuinely occurrent properties. Name the worlds in $\mathscr{W}$ with the help of a suitable index set Δ. The property of a world W_δ ($\delta \in \Delta$) of having in world W_δ a particle with such-and-such a position at such-and-such a time is unique to world W_δ. Since no distinct worlds in $\mathscr{W}$ ever agree on such world indexical properties the proposed working definition of determinism will be vacuously satisfied if such properties are allowed to adjudicate the agreement or disagreement of worlds at a given time.

Our definition of determinism also risks triviality unless we banish overt and covert reference to past and future times. In some sense of property, it is now a property of the pen I am holding that five minutes ago it executed various motions and that five minutes from now it will be at rest on the desk. But such properties are not truly occurrent and like the world indexical properties should not be allowed to decide the agreement of worlds at a given time. The challenge is to distinguish the desired class of occurrent properties without begging the question of determinism. I will simply assume that the challenge can be met.

Even so there are still worries about the meaning of determinism and the adequacy of our definition of it. I will describe two of the worries with the help of a little cracked theology. Imagine that God is perversely energetic in His creation of the physically possible worlds. Specifically, He so loves diversity that He arranges it that the same (non-indexical, truly occurrent) instantaneous state never appears more than once in $\mathcal{W}$. This assures as a byproduct of our working definition that determinism cannot fail whatever else is true about the temporal evolution of the world.

The second example involves a more sacrilegious assumption. Imagine that at the dawn of creation God is fatigued. He is not up to instituting any physical laws in the sense familiar from physics. But because He desires things to go smoothly, He decrees that all physical magnitudes shall be analytic functions of time. If we admit into the instantaneous description of the world not only the instantaneous values of all the basic physical magnitudes but also the instantaneous values of their time derivatives of all orders (surely, all truly instantaneous properties), then the world is automatically Laplacian deterministic over at least some finite interval of time. For by definition, an analytic function can be expanded as a convergent power series; then just plug in the initial value of the function and the values of its time derivatives at the starting time to obtain the values of the function for earlier and later times within the radius of convergence. And even if the instantaneous state description excludes time derivatives of an arbitrary order, analyticity still backs a weakened version of Laplacian determinism since agreement over a finite stretch, no matter how short, on the basic magnitudes will force agreement over a longer stretch.

. When the theology is stripped from these examples what we are left with are questions about how determinism is implemented by physical laws. In both examples the worry is that although the letter of the definition of determinism is satisfied, the spirit is not, since it is not at all evident that the future state is determined, in the intended sense, as a result of the way natural laws guide the unfolding of events. I do not believe that it is fruitful to try to assuage these worries by giving fancier definitions of determinism. Rather, the worries are best dealt with on a case by case basis in terms of the specifics of concrete laws. Towards this end the coming chapters will analyze in some detail a large number of examples drawn from mathematical physics. Even within this circumscribed context it is often difficult to decide whether we have something that deserves to be called genuine determinism. Even at its core the concept of determinism is slippery, and at its outer limits it is altogether too vague to make it worthwhile worrying in the abstract about whether determinism really and truly reigns whenever agreement of worlds at one time forces agreement at another time.

8. DEMOCRACY AND SYMMETRY

The laws of physics, or what have passed for them, have typically displayed a temporally symmetric form of determinism where futuristic and historical determinism stand or fall together. This feature derives from the fact that, until quite recently, all of the fundamental laws of physics were thought to be invariant under time reversal. Note, however, that while time reversal invariance is sufficient for symmetry with respect to futuristic and historical determinism, it is not necessary. The nature of various time symmetries such as time reversal invariance, time translation invariance, periodicity, etc., and their connections to determinism will be discussed in Ch. VII.

Continuous space-time symmetries foster democracy for determinism. If, for example, the laws of nature are invariant under space translation, then for any $W \in \mathcal{W}$, W is deterministic iff every W^s generated from W by space translation is likewise deterministic. General democracy reigns when determinism holds for all members of $\mathcal{W}$ when it holds for any and fails for all when it fails for any. Linear field laws typically display this democratic character. But lest it be thought that determinism requires democracy, I will mention that interesting forms of determinism are compatible with subtle forms of anti-democratic behavior. Let $\mathcal{W}_n^G$ stand for worlds which forever and always contain no more than n point mass particles obeying Newton's laws of gravitation. Then assuming that no collisions occur, $\mathcal{W}_2^G$ and $\mathcal{W}_3^G$ form deterministic collections whereas $\mathcal{W}_5^G$ does not. Or again, let $\mathcal{W}_n^B$ stand for worlds with n equal mass billiard balls obeying the laws of elastic impact. Then for any finite n, $\mathcal{W}_n^B$ is a deterministic collection while $\mathcal{W}_{+\infty}^B$ is not. These matters will be illustrated in Ch. III.

9. NON-LAPLACIAN VARIETIES OF DETERMINISM

According to the Laplacian brand of determinism, the instantaneous state of the world suffices to uniquely fix the state at any other time. Other varieties of determinism can be produced by modifying the types of space-time regions which are determined and which do the determining. Thus, we can say that $W \in \mathcal{W}$ is (R_1, R_2) deterministic just in case for any $W' \in \mathcal{W}$, if W and W' agree on space-time regions of type R_1, then they agree on regions of type R_2. Laplacian determinism simpliciter is (R_1, R_2) determinism with R_1 a time slice and R_2 the rest of space-time; futuristic Laplacian determinism is (R_1, R_2) determinism with R_1 a time slice and R_2 the future of that slice; etc. Close cousins of Laplacian determinism can be obtained by taking the determining region R_1 to be a finite sandwich instead of an infinitely thin slice, the entire past lying below a time slice, etc. These relatives will prove to be useful in discussing determinism for relativistic particle mechanics where strict Laplacian determinism may fail (see Ch. IV).

Laplacian determinism and its close relatives are, to my knowledge, the only varieties which have received attention in the philosophical literature. The explanation cannot be that no other variety is relevant to the analysis of modern science, for giving data on a null surface in relativistic physics is in some respects more natural than giving it on a time slice (see Ch. X). I suspect that the reasons derive from the widespread association of determinism and prediction and the preoccupation with examples drawn from classical physics. But once the illicit nature of the association is revealed and once our horizons are extended beyond the classical, the way is opened for considering non-standard forms of determinism.

10. CHE SARÀ SARÀ

What has been has been; what is is; and just as surely and just as trivially, what will be will be. But the refrain "the future's not ours" is supposed to indicate an inevitability that goes beyond these trivialities. Laplacian determinism entails one kind of non-trivial inevitability: given the way things are now, the future can't be other than it will be, where the 'can't' is the can't of physical impossibility. This is an interesting kind of inevitability but it doesn't quite capture the full sense of uncontrollability and unavoidability of fatalist teachings.

Let us say that an event or state is *X-fated* just in case it occurs in every *X*-possible world. (Alternatively, we might want to say that the event is *X*-fated for an individual *i* just in case that event happens to the individual in every world where *i* (or an *i* counterpart) exists.) Thus, an event is *naturalistically fated* just in case it occurs in every physically possible world. If there are such fated events, then in one clear sense some things are going to happen no matter what — vary the initial conditions as much as you like (within the bounds of physical possibility) and the fated event will nonetheless eventuate. Naturalistic fatalism in this sense neither entails nor is entailed by determinism. Nor is naturalistic fatalism a very controversial or exciting doctrine, being illustrated by the most commonplace of examples. I take it, for instance, that the laws of biology dictate that I am naturalistically fated to die; but I also take it that the particular time and manner of my death are not fated by any of the laws of nature. This is, perhaps, Aristotle's point when he wrote:

> . . . it is necessary that he who lives shall one day die . . . But whether he dies by disease or by violence, is not yet determined, but depends on the happening of something else. (*Meta.* 1027b, 10—14)

Weaker types of natural fate can also be defined. A feature can be said to *asymptotically fated* if it emerges in the limit as $t \to +\infty$ for every $W \in \mathscr{W}$. Such is the hypothesized 'heat death' of the world which may not obtain at any finite time in the future but may emerge in the limit. Or a feature can be said to be *weakly fated* in the actual world $W_{@}$ if it emerges in every $W \in \mathscr{W}$ which has a state that does not depart from the present state of $W_{@}$ by more than some specified degree. What we are obviously moving towards is an association between fatalism and stability: the fated features are the ones which are stable under variations of starting conditions. This is a notion which admits of degrees and the reader is invited to quantify for himself.

Contrary to what is sometimes asserted, perhaps in an attempt to stigmatize, fatalism as applied to human actions need not entail that actions are inefficacious in the objectionable sense that they are 'causally discontinuous' with the future (see Wilson (1955)). Fatalism can allow that our actions do have effects; it is rather that the hand of Fate — as it acts here through the laws of nature — shapes the course of events so that the effects of our actions bring about the fated event. Oedipus was fated to kill his father Laertes and marry his mother Jocasta. All his strivings to avoid his fate were not without effects; indeed, it was these very actions which brought about his fate.

Of course, Oedipus' fate was not induced naturalistically. Still, the above analysis will serve with the appropriate replacement for the collection $\mathscr{W}$ of physically possible worlds. The suggestion is that we construe the workings of super-natural fate in terms of higher laws which are imposed over and above the natural laws. Perhaps these super-natural laws come in the form of decrees of Gods, decrees of Fate, or what-have-you. The resulting set $\tilde{\mathscr{W}}$ of super-naturally possible worlds is then a subset of $\mathscr{W}$. The most extreme version has $\tilde{\mathscr{W}} = \{ W_@ \}$, as would follow if, for example, God necessarily chooses to actualize the best of all possible worlds, with the result that every actual event is super-naturally fated. Leibniz sought to avoid this absolute metaphysical fatalism, though his principle of sufficient reason pushes him towards it.

There is a long tradition in philosophy which seeks to prove that the laws of logic suffice to establish fatalism for human actions (see Taylor (1983)). From the perspective of our analysis of fatalism, this tradition is opaque, for it would seem that the laws of logic narrow the collection $\mathscr{W}$ of physically possible worlds not one wit, each of the members of $\mathscr{W}$ having been antecedently assumed to be logically possible. So either we have failed to appreciate the relevant sense in which human actions are fated, or we have misunderstood what the 'laws of logic' involve, or the classical proofs of fatalism are so much hocus-pocus. The actual situation is, I think, a mixture of all three (the main ingredient being the third), but this is not the place to try to disentangle the mess. The Idealists contend that physical necessity is but dimly perceived logical necessity (see Ch. V), with the results that the set $\mathscr{W}$ of physically possible worlds is coextensive with the set of logically possible worlds and, hence, that naturalistic and logical fate are the same. But these results seem to make fatalism harder, not easier, to secure.

11. DETERMINISTIC THEORIES

Positivism and logical empiricism promoted a fear of the ontological and a flight towards the linguistic. Thus, it is not surprising to find that many philosophical discussions of determinism are couched in terms of theories, construed as linguistic entities. But since determinism is a doctrine about the nature of the world, no problem is avoided by this linguistic detour; for to be adequate, a definition of determinism in terms of theories must guarantee that the axioms of the theory express laws of nature and that these laws have just the deterministic property required in the possible worlds definition. It may be, however, that while no problems are avoided, gains in understanding are made by taking the linguistic route. For instance, it may be that the concept of laws of nature is inextricably bound up with scientific theorizing.[5] Or it may be that we can get a firmer grip on how ontological determinism operates by clarifying the concept of a deterministic theory and then studying examples of such theories.

In what is almost standard usage, philosophers identify theories with deductively closed sets of sentences of some formal language. E. Nagel (1953, 1961), Smart (1968), and others have recommended a syntactical characterization of determinism for such theories. Roughly, a theory T is deterministic just in case, given the state description $s(t_1)$ at any time t_1, the state description $s(t_2)$ at any other time t_2 is deducible from T. Montague (1974) noted that for the kinds of formal languages commonly used, there is a difficulty in giving this definition a literal reading. A 'state description' is, presumably, a sentence of the language of T, but while there may well be a non-denumerable infinity of physically possible states of a system, the standard formal languages contain only a denumerable number of sentences.

Two reactions are possible. We can resort to infinitary languages which have the requisite expressive power. However, this increased power may be purchased at the expense of some ugly logical features, e.g., completeness may fail so that the relation of deducibility may not after all be appropriate for characterizing determinism for theories. The other approach, explored by Montague, is to stick with standard languages but to switch from a syntactic to a semantic analysis which mirrors the possible worlds definition using models of T as the counterparts of possible worlds; roughly, T is deterministic just in case for any pair of models of T, if they agree at one time then they agree at all times. Though the idea is the same, there is a gain in working through the formal details of how the concepts are implemented for concrete T's. The interested reader is referred to Montague's brilliant pioneering work.[6]

The formal-systems approach will not play much of a role in my discussion of substantive issues of determinism in modern physics. Most of the putative laws of physics take the form of differential equations for which questions of determinism principally involve existence and uniqueness properties of solutions, and these properties can be discussed with as much rigor as is ever needed without having to resort to formal systems. If philosophers had spent less time trying to achieve for determinism the superficial 'precision' afforded by formal symbolic notation and had spent more time studying the content of physical theories they might have confronted the truly fascinating substantive challenges that determinism must face in classical and relativistic physics. Most philosophers pay lip service to Carl Hempel's remark that there is no real gain in clarity and precision to be had by translating 'A man crossed the street' into 'There exists a man m, a street s, and a time t such that . . .' But many seem to cling to the notion that an advance is to be achieved by applying really powerful formal machinery. Good luck to them.

12. CONCLUSION

We have barely begun and already we are in very deep waters. Space, time and space-time; laws, theories, and formal systems; symmetries and invariances; cause and effect; prediction, instability, and randomness; materialism and physicalism — these are some of the concepts we have encountered in trying to get no more than a preliminary fix on determinism. This is already enough to make strong the suspicion that a real understanding of determinism cannot be achieved without simultaneously constructing a comprehensive philosophy of science. Since I have no such comprehensive view to offer, I approach the task I have set myself with humility. And also with the cowardly resolve to issue disclaimers whenever the going gets too rough. But even in a cowardly approach, determinism wins our unceasing admiration in forcing to the surface many of the more important and intriguing issues in the length and breadth of the philosophy of science.

NOTES

[1] If the only restrictions on the motions of the particles are that they move rectilinearly between collisions and behave like elastic billiard balls in collisions, then the motions are demonstrably non-deterministic except for some very special kinds of collisions; see Ch. III below.

[2] Laplace (1820), Preface; translation from E. Nagel (1961), pp. 281—282. Laplace seems to have given the wrong initial data problem for Newtonian gravitation; see Ch. III.

[3] There is an indirect but important connection between determinism and simplicity. Determinism (as I formulate it below) is a property of laws of nature, and simplicity is one of the features used to separate lawful from non-lawful regularities (see Ch. V).

[4] This assumes that the allowed instantaneous states are the same at every moment of time, an assumption which may fail if the laws are not time translation invariant (see Ch. VII). If the laws are time translation invariant, only the interval $t_2 - t_1$ matters and we can write $s(t_2) = F(s(t_1), t_2 - t_1)$. This is the sense in which the new Russell function need not involve time explicitly.

[5] This is the theme of most of the recent attempts to characterize natural laws; see Ch. V.

[6] Note, however, that Montague's approach is not without its potential pitfalls. Since any one of the standard formal systems of the type Montague studies is capable of representing at most a countable number of magnitudes, the possibility that there are an uncountable number of distinct physical magnitudes which interact with one another so as to produce a deterministic evolution has to be ignored. Russell's notion of determinism can be rehabilitated by requiring that there is a function which is definable in the formal system and which expresses the state at t in terms of t, t_0, and the state at t_0. Montague shows that for what he calls predicative theories this requirement is strictly stronger than determinism.

SUGGESTED READINGS FOR CHAPTER II

A fair sampling of how philosophers have sought to analyze the meaning of determinism is to be gained from Chs. 1 and 2 of Popper's (1982) *The Open Universe*, Russell's (1953) "On the Notion of Cause," E. Nagel's (1953) "The Causal Character of Modern Physical Theory," and Montague's (1974) "Deterministic Theories." The chapter on "Fate" from Taylor's (1983) *Metaphysics* and Cahn's (1967) *Fate, Logic, and Time* contain information on the standard philosophical views of fatalism.

CHAPTER III

DETERMINISM IN CLASSICAL PHYSICS

> All events, even those which on account of their insignificance do not seem to follow from the great laws of nature, are a result of it just as necessarily as the revolutions of the sun. In ignorance of the ties which unite such events to the entire system of the universe, they have been made to depend upon final causes or upon hazard, according as they occur and are repeated with regularity, or appear without regard to order; but these imaginary causes have gradually receded with the widening bounds of knowledge and disappear entirely before sound philosophy, which sees in them only the expression of our ignorance of the true causes.
>
> (P. S. Laplace, *A Philosophical Essay on Probabilities*)

This passage has been taken as a classic statement of determinism, and if it is then it is easy to appreciate how determinism came to occupy such an exalted status: if the only alternatives to determinism are final causes (e.g., divine intervention) and hazard (e.g., accident or chance), then determinism is attractive as an *a priori* truth or a methodological imperative of scientific inquiry. But some care is needed here, as already hinted in Ch. II; for Laplacian determinism as I have proposed to understand it need not be true even though all events are subject to laws that leave no room for divine intervention or accident. Classical physics would seem to be a poor choice of hunting grounds for such examples since, as we all know, the laws of classical physics are deterministic in the Laplacian sense. We know no such thing, at least if knowledge implies truth.

1. CLASSICAL WORLDS

The initial setting for the doctrine of determinism was what I called the classical world picture. It is time to be more specific about how that picture is composed. There are three features which require special emphasis. (1) All the members of the set $\mathcal{W}$ of physically possible classical worlds are assumed to have a common space-time back-ground. This common space-time is the canvas on which the possible worlds are painted. The details of the structure of the canvas will turn out to be as crucial to the success or failure as what is painted on it: *too little structure of the right kind or too much structure of the wrong kind and determinism will never succeed no matter how furiously or cleverly we paint.* This important but largely unappreciated moral will be drawn in detail in this and succeeding chapters, but for now I will reemphasize only one element of classical space-time structure. Namely, (2) the four-dimensional space-time canvas is ruled by a family of three-dimensional hypersurfaces called the planes of absolute simultaneity; two events are simultaneous just in case they lie on the same plane. (3) The canvas is filled in by specifying the values of a collection of physical magnitudes, each of which is assumed to be a point valued quantity.

If, for sake of definiteness, we think of the physical magnitudes as geometric object fields on space-time, then classical worlds can be presented in the form of a triple $\langle M, \{G_\alpha\}, \{P_\beta\}\rangle$, $\alpha \in \mathcal{A}$, $\beta \in \mathcal{B}$ ($\mathcal{A}, \mathcal{B}$ index sets) where M is the space-time manifold (usually assumed to be $\mathbb{R}^4$), the G_α are geometric object fields on M characterizing the struc-ture of space-time (including, of course, the simultaneity structure (2)), and the P_β are geometric object fields characterizing the physical contents of space-time. In keeping with (1), M and the G_α are common to all members of $\mathcal{W}$ while the P_β vary from world to world. Agreement of two worlds $\langle M, \{G_\alpha\}, \{P_\beta\}\rangle$ and $\langle M, \{G_\alpha\}, \{P'_\beta\}\rangle$ at a given time means agreement on a plane of absolute simultaneity of the values of the physical magnitudes.[1]

Modern physics contradicts or challenges each of the assumptions (1)–(3). The special theory of relativity contradicts (2); the general theory contradicts (1); and, according to some interpretations of quan-tum physics, quantum theory undermines (3). What happens to the doctrine of determinism when one or more of the props of the classical worlds is kicked out will have to be discussed in detail in later chapters. For the moment, let us assume that the props are secure. What is surprising is that even with their support, classical worlds prove to be an unfriendly environment for any form of Laplacian determinism. To the extent that determinism passes the Scylla of triviality, it appears to run a ground on the Charybdis of falsity. In Ch. II we viewed the Scylla. We must now face the Charybdis.

2. THE APPARENT FAILURE OF DETERMINISM IN LEIBNIZIAN PHYSICS

Some of the versions of Leibniz's multi-faceted principle of sufficient reason either entail or presuppose determinism. And yet, as Howard Stein (1977) has shown, Leibniz's views on the nature of space and time seem to preclude any interesting form of Laplacian determinism.

Let us recall Leibniz's version of the space-time structure of classical worlds. In accordance with the characterization of Sec. 1 above he agreed that there is an absolute notion of coexistence or simultaneity. And like all 17th century natural philosophers, he assumed that the instantaneous three-spaces have a Euclidean $\mathbb{E}^3$ structure and, further, that there is a well-defined sense of duration or temporal distance for non-simultaneous events. Finally — and this is the crucial point — he held that these elements completely exhaust the structure of space-time. The symmetry mapping of this Leibnizian space-time can be presented in the following form. Let x^α, $\alpha = 1, 2, 3$, stand for a Euclidean coordinate system; and let t stand for absolute time, i.e., $t: M \to \mathbb{R}$ is such that its level surfaces coincide with the planes of absolute simultaneity and the intervals $|t_1 - t_2|$ give the duration between the events e_1 and e_2 lying respectively on the planes $t = t_1$ and $t = t_2$. Then the symmetry maps have the form:

$$(L) \qquad x^\alpha \to x'^\alpha = R^\alpha_\beta(t)x^\alpha + a^\alpha(t)$$

$$t \to t' = t + b$$

where b is a constant, $a^\alpha(t)$ is an arbitrary smooth function of t, and $R^\alpha_\beta(t)$ is a time dependent orthogonal matrix. In words, the structure preserving maps of Leibnizian space-time onto itself are time translations and (possibly) time dependent Euclidean spatial translations and rotations.

On this canvas Leibniz painted a plenum of matter; but for ease of illustration it suffices to consider more sparsely populated worlds containing, say, three particles.

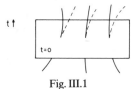

Fig. III.1

Such a world is completely described by drawing in the world lines of the three particles (the solid curves of Fig. III.1). Consider how the mappings (L) act on the particle histories. From among the mappings we can choose one which reduces to the identity for all $t \leqslant 0$ but which is non-trivial for $t > 0$. It leaves fixed the entire past history of the particles but changes their future behavior, as indicated schematically by the dashed lines. Thus, if the Leibnizian laws of motion satisfy the demand that the space-time symmetries are also symmetries of the laws (i.e., carry a physically possible history to another physically possible history), then we have a violation of even the weakest form of Laplacian determinism; for given any physically possible history of the particle trio, there will be another physically possible history which agrees with the first for all past times but disagrees in the future.

The announced demand is eminently reasonable, as is the stronger demand that the space-time symmetries and the symmetries of the laws of motion coincide. If the symmetries of the laws were more inclusive than the symmetries of the space-time, then the space-time would contain more structure than is needed to support the laws and Occam's razor would slice it away. On the other hand, the symmetries of the laws should be at least as wide as the symmetries of the space-time; for if the laws allow one history but not another, then those histories cannot be connected by a space-time symmetry — otherwise, there would be no way to express the difference between the allowed and the prohibited histories in terms of the behavior of physical magnitudes on the space-time canvas. Technically, the underlying assumption is that the set of models of the laws are closed under automorphisms of the space-time background, i.e., if $\langle M, \{G_\alpha\}, \{P_\beta\} \rangle$ is a model and d is a diffeomorphism of M onto itself such that $d^*G_\alpha = G_\alpha$ for each α, then $\langle M, \{G_\alpha\}, \{d^*P_\beta\} \rangle$ is also a model, where d^* denotes the 'drag along' by d. Conceivably, a theory of motion could postulate different lawlike behaviors in different space-time regions. But such a difference would be grounds for distinguishing the regions in terms of absolute structure; that is, for adding, if necessary, elements to $\{G_\alpha\}$ so that the regions in question are not connected by a space-time symmetry. And it seems that in this manner our assumption can always be vouchsafed (see Earman (1986) for details).

There are two ways to bridge the abyss which has opened between the vision of determinism and its fulfillment. One relies on the reinterpretation of Leibnizian space-time, the other on an enrichment of it. The two moves will be briefly reviewed in turn.

3. LEIBNIZ'S RESPONSE

Leibniz would have welcomed this challenge as an opportunity to expose the Achilles' heel of the Newtonian conception of the space-time manifold, or as he put it in the famous correspondence with Samuel Clarke, to "confute the fancy of those who take space to be a substance."

Note that the transformations (L) preserve all relative particle quantities such as relative distances, relative velocities, relative accelerations, etc. According to Leibniz, facts about the values of these relative quantities exhaust the factual content of the physical world. Thus, the 'two' world histories pictured in Fig. III.1 do not really correspond to objectively different worlds but only to different descriptions of the same world. Consequently, the alleged violation of determinism is only an illusion due to the descriptive fluff packed into our presentation of classical space-time worlds.

Leibniz's position here does not result from a question-begging desire to save determinism, but is arrived at by an independent route that passes through his meta-physics and his metaphysics. According to the former, which owes much to Descartes and Huygens, all motion must be analyzed as the relative motion of bodies. According to the latter, there would be a violation of the principle of sufficient reason if Fig. III.1 did illustrate objectively different world histories; for in deciding which of the two worlds to actualize, God would find Himself in a Buridan's ass situation, unable to choose between two worlds which are not separated by any properties that provide sufficient grounds for choice. As Leibniz put it in the third letter to Clarke:

I say then, that if space was an absolute being, there would something happen for which it would be impossible there should be sufficient reason. Which is against my axiom. And I prove it thus. Space is something absolutely uniform; and, without the things placed in it, one point of space does not absolutely differ in any respect whatsoever from another point of space. Now from hence it follows, (supposing space to be something in itself, besides the order of bodies among themselves,) that 'tis impossible there should be a reason, why God, preserving the same situations of bodies among themselves, should have placed them in space after one certain particular manner, and not otherwise; why every thing was not placed the quite contrary way, for instance, by changing East into West. But if space is nothing else, but that order or relation; and is nothing at all without bodies, but the possibility of placing them; then those two states, the one such as now is, the other supposed to be quite the contrary way, would not all differ from one another. Their difference therefore is only to be found in the chimerical supposition of the reality of space in itself. But in truth the one would exactly be the same thing as the other, they being absolute indiscernible; and consequently there is no room to enquire after a reason of the preference of the one to the other. (Alexander (1956), p. 26)

The philosophical reaction to Leibniz's critique has tended to divide: those who share with him the notion that all motion is relative bodily motion are naturally sympathetic while those who are impressed by the fact that neither classcial *nor* relativistic physics supports this notion are less sympathetic. What both sides have failed to see (and what Leibniz himself was not clear about) is that the issue of relationism is not equivalent to the key issue Leibniz raises about our mode of presentation of space-time worlds. As I read it, his objection is first and foremost to the view that space-time is a kind of substance or container which exists over and above the events it houses. The objection can be stated in a form that is independent of the intertwined questions of whether all motion is the relative motion of bodies and what goes into the G_α. Let d be any diffeomorphism of the space-time manifold M onto itself. For fields G_α and P_β, d induces new ones d^*G and d^*P respectively (the fields 'dragged along' by d). Any two models $\langle M, \{G_\alpha\}, \{P_\beta\}\rangle$ and $\langle M, \{d^*G_\alpha\}, \{d^*P_\beta\}\rangle$ related in this way are by Leibniz's lights just different modes of presentation of the same physical reality. And this is so even if the structure of space-time, as specified by $\{G_\alpha\}$, falsifies the slogan "All motion is the relative motion of bodies," as it is falsified for orthodox Newtonian and special relativistic space-time, both of which contain inertial structure that permits the definition of absolute or invariant dynamical quantities, such as acceleration, which are *not* relative particle quantities (see below). Further, this is so even if space (or space-time) is not "absolutely uniform" but is, say, variably curved; for this curvature is represented by some appropriate object in $\{G_\alpha\}$ and is dragged along by d along with everything else so that again the original model and its image model "do not at all differ from one another" and are "absolutely indiscernible."

Thus, on my interpretation the essence of Leibniz's objection is to treating points and regions of M as real existents, as substances in the proper logical sense of objects of predication. There is a quick and cheap way to reform our presentation of space-time models so as to escape the objection; namely, take equivalence classes of "indiscernible" models and declare that each class corresponds to a single Leibnizian world. The more interesting challenge is to start from the other end and give a direct and intrinsic characterization of the Leibnizian worlds and then show that the members of an equivalence class of ordinary models arise as different but equivalent representations of the same intrinsic reality. For someone like myself who is not a relationist and who does not believe that all motion is the relative motion of bodies, the challenge takes the form of erasing the underlying manifold M of space-time points while keeping the non-relational structure of space-time, a kind of Cheshire cat trick.[2]

Whatever the ultimate decision on the ontological status of space-time, there remains the problem of what geometric structure G_α and physical magnitudes P_β are needed in an adequate theory of motion. And here the weight of evidence goes strongly against Leibniz. From Galileo to Newton to Einstein, every successful theory of motion makes use of physical quantities which cannot be reduced to relative particle quantities. This opens up a new avenue along which determinism can move; for in order to have well-defined absolute, or non-relative, quantities of motion, the structure of Leibnizian space-time must be beefed up. Consequently, the symmetries (L) must be cut down. Such a cutting down may also cut down the counterexamples to determinism.

4. NEWTONIAN SPACE-TIME

Newton's space-time canvas is much more complex than Leibniz's. In addition to simultaneity, duration, and Euclidean space structure, it also contains a preferred family of motions, called inertial frames, and a distinguished family member called absolute space. The addition of the inertial structure makes into well-defined quantities ones which are not well-defined or invariant in Leibnizian space-time — in particular, the instantaneous (non-relative) acceleration of a particle — and it linearizes the space-time symmetries to form the familiar Galilean transformations:

$$(G) \qquad x^a \to x'^a = R^a_\beta x^\beta + v^a t + c^a$$

$$t \to t' = t + b$$

where v^a and c^a are constants and R^a_β is now a constant orthogonal matrix.

The further addition of absolute space in the sense of a distinguished inertial frame makes (non-relative) velocity as well as acceleration a well-defined dynamical quantity and reduces the space-time symmetries to

$$(N) \qquad x^a \to x'^a = R^a_\beta x^\beta + c^a$$

$$t \to t' = t + b$$

The objection to full-blown Newtonian space-time, with this form of absolute space, as a setting for mechanics is that it violates the principle enunciated in Sec. 2 above connecting symmetries of space-time and symmetries of laws; for the Newtonian laws of motion are invariant under (G). It is perfectly conceivable, however, that additional laws might break the Galilean invariance, necessitating the introduction of additional space-time structure and narrowing (G) to (N); in fact, it was thought in the 19th century that the laws of optics and electro-magnetism did just that. I will return to this matter in Sec. 14 below, but for the moment it can be ignored since the addition of the inertial structure to Leibnizian space-time is in itself sufficient to block the argument which threatened Leibnizian determinism; for any member of (G) which reduces to the identity for any finite interval of time, no matter how short, is the identity map everywhere. Note, however, that without the help of absolute space there are limitations to Newtonian determinism. For example, it is not possible to write a law which allows a scalar quantity Φ to vary in space at a fixed moment of time and which determines the future values of Φ from its initial value $\Phi(x, 0)$, $-\infty < x < +\infty$, at $t = 0$. For the law has to be Galilean invariant so that the application of Galilean transformation to any solution must produce a new solution. Choose the transformation so that it is the identity for $t = 0$ but not for later times. Since the initial data are preserved, we then will have two solutions which agree at $t = 0$ but differ in the future.

5. NEWTONIAN PARTICLE MECHANICS

Because the above considerations have been very abstract, it is useful to have before us some concrete examples of determinism triumphant. Since Laplace's espousal of determinism was prompted by his reflec-tions on Newtonian celestial mechanics, it would be natural to look there for the desired example, but actually it turns out to be cleaner to envision a force law different from Newton's $1/r^2$ law.

Consider N point masses m_i $(m_i > 0)$, $i = 1, 2, 3, \ldots, N$, and suppose that they attract each other in pairs with a force which acts along the line joining them and which is proportional to the product of their masses and the distance separating them.[3] Combining this force law with Newton's second law of motion yields:

$$(\text{III.1}) \quad m_k \ddot{r}_k = \sum_{\substack{j \\ j \neq k}} C m_j m_k (r_j - r_k) \qquad (C = \text{positive constant})$$

It is always possible to find an inertial system in which the center of mass is at rest at the spatial origin. In such a system, the equations (III.1) decouple and as a result, the initial value problem with given initial data

(III.2) $r_k(0) = \mathring{r}_k$, $v_k(0) \equiv \dot{r}_k(0) = \mathring{v}_k$

not only has a unique solution, but the general solution can be written down in closed form. Every physically possible history of the system is thus comprehended in a single analytic formula, and the possible pasts and possible futures of the system are, in Laplace's words, present before our eyes.

For Newtonian gravitation, the equation of motion is

(III.3) $m_k \ddot{r}_k = \sum_{\substack{j \\ j \neq k}} Gm_j m_k (r_j - r_k)/r_{jk}^3$ $(r_{jk} \equiv |r_j - r_k|)$

The initial value problem has a unique solution, at least *locally* in time. If all $r_{ij} \neq 0$, $i \neq j$, at $t = 0$, there exist unique functions r_k of t and a time interval (t_1, t_2) such that for any $t \in (t_1, t_2)$ (III.3) holds and for $t = 0$ (III.2) holds. When $N \geqslant 3$, there are initial conditions for which t_1 or t_2 (or both) are finite. If the solution cannot be extended as a smooth function of t to $t_2 = +\infty$ and $t_1 = -\infty$, the solution is said to be *singular*.

If all such singularities were due to collisions of two or more of the point particles, we could affirm a qualified doctrine of determinism:

(Q)   Barring collisions, Newtonian gravitational theory of point mass particles is Laplacian deterministic.

And we can make (Q) sound more impressive by adding that the antecedent is almost always satisfied, for it is known that the set of initial conditions which lead in a finite time to collisions is of (Lebesgue) measure zero (Saari (1973)).

There are, however, some caveats about (Q). First, measure zero need not imply either insignificant or ignorable. We would not judge the set of initial conditions giving rise to collisions to be insignificant if, for example, it proved to be dense within the set of states that eventuate in strong interactions (in some appropriate sense) among the particles. Nor would we regard the measure zero set as ignorable if it loomed large within the range of cases we regard, for whatever reason, to be physically interesting. To illustrate, take the case of $N = 2$. Here it is easy to see that the set of states leading to collisions has measure zero since a collision cannot occur unless the angular momentum is zero and since for $N = 2$ the set of zero angular momentum states has measure zero. But if we are interested in zero angular momentum states, then collisions loom large — indeed, for $N = 2$ such states always lead to collisions.

The second and more important caveat is that (Q) may be false! Define $r(t) \equiv \min(r_{ij}(t))$, $i \neq j$. Then if (t_1, t_2) is the maximal interval for which the solution exists, t_2 is finite iff $r \to 0$ as $t \to t_2^-$, and similarly, t_1 is finite iff $r \to 0$ as $t \to t_1^+$. Further, for $N \leqslant 3$, $r \to 0$ iff there is a collision. But for $N \geqslant 4$ it is an open question as to whether or not $r \to 0$ implies a collision, though the evidence now available indicates a negative answer (see Sec. 7 below). How might the implication fail? One can first try to imagine that the occupant of the role of the minimum r_{ij} switches around and around. But since there are only a finite number of particles, at least one of the potential occupiers, say r_{34}, must actually occupy the role an infinite number of times as (say) t_2 is approached. Thus, we are forced to imagine an oscillatory behavior in r_{34} with lim inf $r_{34} = 0$ but lim sup $r_{34} > 0$. Such wiggling is used in constructing anomalous solutions, as we will see below in Sec. 7. But note that even if r_{34} does go to zero there need not of mathematical necessity be a collision in the proper sense that the position vectors of particles 3 and 4 both approach the same fixed point in space. For it is mathematically possible that these particles accelerate themselves off the space-time manifold and cease to exist at t_2. And a theorem of Sperling (1970) shows that such unbounded behavior must occur in non-collision singularities, should they exist.

It is shocking that determinism may break down for the very case which was supposed to serve as a paradigm example of determinism at work. But worse still, reflecting on the way determinism might break down in this case leads to a general worry about how determinism could ever be securely established in classical worlds.

Before examining the reasons behind this paradoxical worry, let us take note of the somewhat less paradoxical opposite side of this coin. We will see in Ch. VI that to the extent that determinism holds in this case, its course can be traced out by a dumb ($\neq$ stupid) digital computer, if the initial data are computable. Thus, although Laplace was overly optimistic in one way he was overly pessimistic in another; for to the extent that his demon is possible, it need not remain "infinitely remote" but can be instantiated by an uncreative mechanical calculator.

6. DETERMINISM AT BAY

Only a little reflection on some of the commonplaces of classical physics is needed to switch the *Gestalt* of determinism safely and smoothly at work in Newtonian worlds to puzzlement about how Laplacian determinism could possibly be true. The first commonplace is that it is hopeless to try to establish determinism for a system which is not closed to outside influences. Trying to determine the weather in Minneapolis tomorrow from even the most precise meteorological data today in Minneapolis is a thankless task since tomorrow's weather can be influenced by what is now happening in North Dakota and Wisconsin (Fig. III.2). Two remedies may be contemplated. This first is to erect imaginary boundary walls (W_1 and W_2 of Fig. III.3) to record the incoming influences as they penetrate the boundaries of the system. This gives rise to a non-Laplacian initial-boundary value problem: given the appropriate initial data on S and the appropriate boundary data on $W_1 \cup W_2$, determine the state in the interior region R. For field theories where there is action by contact such initial-boundary value problems are often well posed; a successful example will be considered below in Sec. 11. But success cannot be expected for action at a distance theories where effects are transmitted without leaving any traces on the intervening spaces. Furthermore, even when the initial-boundary value approach is successful, the success relies on a departure from pure Laplacian determinism by requiring a specification of future data. I therefore turn to a second remedy which seeks to preserve Laplacian determinism in its pristine form.

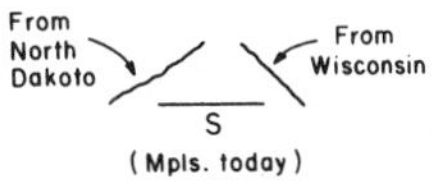

Fig. III.2 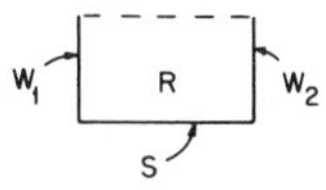   Fig. III.3

The outside influences coming from the Dakotas and Wisconsin can be co-opted by extending the boundaries of the system to take in the hinterlands. For practical purposes, a finite extension of the original initial data surface S may suffice for a pretty good determination of tomorrow's weather. But in a spatially infinite universe, S must be extended to infinity in all directions to make sure that the co-option is complete enough to rule out any possibility of a nasty surprise coming from without. In this way we are driven from the local form of Laplacian determinism to the global form.

The second commonplace threatens even the global form. The laws of classical physics place no limitations on the velocity at which causal signals can propagate. This fact is intimately related to the structure of Newtonian space-time. Without absolute space, velocity is not a Newtonian invariant; whatever the finite value of a particle velocity as measured in one inertial frame, there will always be another inertial frame in which the value is as large as you like. Thus, no law of motion invariant under the Galilean transformations can entail the existence of a fixed finite bound on particle velocity. An infinite velocity is, however, an invariant concept within the Galilean group, and this in turn leads back to the justification for absolute simultaneity: distant clocks can, in theory, be brought into absolute synchronization by means of a sequence of signals whose velocities tend towards infinity.

Signals with actually infinite velocities will be considered a little later, but for the moment it is sufficient to contemplate particle or wave motions where the velocities of propagation are everywhere finite but unbounded. Fig. III.4 illustrates the space-time history α of a particle with velocity $|\dot{x}(t)| < \infty$ for all t but with finite 'escape time' $t^* =$ high noon on April 1, 1988.

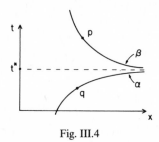

Fig. III.4

As $t \to t^*$ from below, $x(t) \to \infty$ and the particle disappears from the universe, even though $|\dot{x}(t)| < \infty$. The curve β is the temporal mirror image of and it represents the history of a particle which springs the April Fool's joke by appearing from spatial infinity. Thus, in Newtonian space-time the co-option strategy appears to be doomed to failure, for even if the system is extended to include the entire universe, it is not automatically 'closed' in the operative sense to outside influences.[4]

Please do not complain that we never have observed such disturbing disappearing and appearing acts and that, by induction, it is reasonable to expect that we never will. Determinism is a doctrine not just about the actual world but about all physically possible worlds. So even if we can safely employ induction to conclude that no such particles are actual, Laplacian determinism is still threatened if such processes are physically possible. The possibility of $\beta(\alpha)$ is a *prima facie* insult to futuristic (historical) determinism since $\beta(\alpha)$ influences points to the future (past) of $t = t^*$, e.g. $p(q)$, but does not register on $t = t^*$ and so leaves no initial data which can be projected into the future (past).

The threat can be restated by borrowing a concept used extensively in the discussion of relativistic determinism. Let S be a global or local time slice (here, a plane of absolute Newtonian simultaneity or a portion thereof). The *future domain of dependence* $D^+(S)$ of S is to consist of all points p of space-time such that (i) p lies to the future of S and (ii) the state at p depends only on the state on S. The *past domain of dependence* $D^-(S)$ of S is defined analogously. And the *total domain of dependence* $D(S)$ is then the union $D^+(S) \cup D^-(S)$. How to interpret the crucial clause (ii) turns on assumptions about the physics of the situation, but this much seems clear: $p \notin D^+(S)$ (respectively, $D^-(S)$) if there is a space-time curve, representing a physically possible causal signal, which passes through p but which never meets S no matter how far it is extended into the past (respectively, the future). The point of the preceding paragraph can now be restated thusly: Whatever the choice of S in Newtonian space-time, domains of dependence are trivial, for $D(S) = D^+(S) = D^-(S) = S$. Laplacian determinism not only doesn't get to first base, it never even has the chance to come out of the on deck circle! In relativistic space-times, as we will see in the next chapter, determinism at least can be brought to bat in that domains of dependence extend non-trivially.

7. DETERMINISM AT SEA

The threat to determinism is, so far, only a *prima facie* one. To make it palpable, it must be shown that physically possible force functions can generate the kind of behavior picture in Fig. III.4. And more, it must also be shown that the sources which generate the forces either themselves escape contact with $t = t^*$ or else that their behavior at t^* does not code up enough information to make a unique determination of the past and future.

Newtonian gravitational theory of point mass particles provides a relevant example. Mather and McGehee (1975) studied a system of four point mass particles moving colinearly under their mutual Newtonian gravitational forces. Particles 3 and 4 approach one another ever more closely, giving up potential energy in the process. Some of this energy is used to accelerate 3 and 4 and some of it is transferred to particle 1 by means of particle 2 which bounces back and forth between 1 and 3 (see Fig. III.5). Collision singularities are involved, but for the binary collisions the solutions can be extended in a physically reasonable way on the model of elastic bounces. Using this device, Mather and McGehee establish that the solution can become un-bounded in a finite time t^*: as $t \to t^*$ from below, $x_1(t) \to -\infty$, $x_3(t)$, $x_4(t) \to +\infty$, while $x_2(t)$ executes an infinite number of bounces between particles 1 and 3. Since the laws of Newtonian gravitation are invariant under time reversal we can invert the Mather-McGehee scenario to produce a solution which insults futuristic determinism by presenting an empty universe up to t^* but thereafter having four particles, three of which appear from spatial infinity and the other of which oscillates infinitely back and forth.

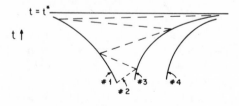

Fig. III.5

Perhaps the problem for determinism is due to collisions. If so we could retreat to the qualified form (Q) of determinism asserting that, barring collisions, Newtonian gravitational theory of point mass parti-cles is Laplacian deterministic. However, it now seems that this retreat does not take us onto safe ground. In a recent article, Gerver (1984) presents a model with five coplanar point masses that never collide. The messenger particle 5 shuttles around the triangle, picking up energy from particle 1 and transferring part of it to 2, 3, and 4, with the result that the triangle expands with each round trip of the messenger (see Fig. III.6). Gerver makes it plausible that the speed of the messenger and the rate of expansion of the triangle can be arranged so that within a finite time the messenger completes an infinite number of round trips while the triangle becomes infinitely large. The details of a rigorous proof remain to be given.

Fig. III.6

It is known that for $N = 4$ the set of initial conditions which could potentially eventuate in a noncollision singularity has measure zero (see Saari (1977)), essentially because, as in the Mather-McGehee construction, all four particles must approach a fixed line in space. But as argued in Sec. 5, measure zero does not necessarily mean insignificant or ignorable, and, moreover, cases for $N > 4$ remain to be settled.

It is not immediately clear to what extent singular but noncollision solutions, should they exist, would undermine Newtonian determinism; for it is not obvious under what conditions such solutions can be joined onto a normal solution. In the case of the heat equation, to be studied below in Sec. 10, the existence of a single solution which is null before $t = 0$ but non-null afterwards is sufficient to completely destroy futuristic determinism since, by the linearity of the equation, the self-exciting solution can be added onto any other solution to produce a new solution.

8. LIFE RAFTS

The true believer in determinism will be undaunted by the examples of Mather-McGehee and Gerver. There is, he will contend, only an apparent failure of determinism, the false appearance being due to considering a space of solutions that is too large in the sense that it encompasses solutions that are not genuinely physically possible; and once these impostors are rooted out, the triumph of determinism will again become apparent. That is the general strategy. Three concrete suggestions for implementing it come to mind.

(i) *Impose boundary conditions at infinity.* By the imposition of appropriate boundary conditions at spatial infinity we can rule out influences coming from or disappearing to God knows where. This achieves by fiat what the laws of motion were supposed to achieve on their own. Given the present state of the universe, the laws determine the future state — as long as there are no rude surprises. Boundary conditions at infinity are just a way of asserting that rude surprises will not be counteranced.

(ii) *Add additional laws.* The escape solutions discussed in the preceding section appear to violate conservation of mass and momentum, so in so far as conservation principles are sacred, the escape solutions are physically impossible. Distinguish two principles of conservation of mass: (C1) particle world lines do not have beginning or end points and mass is constant along a world line, and (C2) for all time t_1 and t_2, the total mass at t_1 = the total mass at t_2. (C1), I claim, is a fundamental principle of classical physics, and it is satisfied even in the anomalous escape solutions. Further, if the laws of motion do not allow escape solutions, then (C1) entails (C2). Some people have been misled into thinking that (C2) is a basic law of classical physics because they have not recognized the possibility of escape solutions.

A similar response is to be made to the invocation of conservation of momentum. Given that a system is closed and that the interactions among the particles satisfy certain restrictions, we can prove conservation of momentum as a theorem. But there is not the ghost of a hope of proving or securing conservation of momentum if the system is open. And the question here is precisely that of whether the universe as a whole is an open system.

(iii) *Object to the idealization of point mass particles.* There are three responses to the objection. (a) Idealizations are always involved in science, and this idealization of point mass particles moving under their mutual gravitational forces was supposed to provide the paradigm of Laplacian determinism at work. So the objection is both querulous and self-defeating. (b) Remove the idealization and consider corpulent particles. You must then say what happens when a collision takes place. Classical physics suggests that we impose laws of elastic impact. But binary collisions of unequal mass particles in two or more spatial dimensions or triple collisions of unequal masses in one spatial dimension are generally non-deterministic. (c) Consider binary collisions of equal mass particles in one spatial dimension. Each collision is determinstic. But with enough particles anomolous non-deterministic solutions can be created, as we will now see.

9. INFINITE BILLIARDS

Consider a system of billiard balls strung out in (two-dimensional) space as shown in Fig. III.7. The balls are assumed to interact only by

Fig. III.7

contact and then according to the Newtonian laws for perfectly elastic bodies. If for $t < t^*$ all the balls are at rest, then barring outside intervention, the balls will remain at rest for $t > t^*$. If, however, the system is infinitely expanded, letting the number n of balls increase without limit, non-uniqueness of future behavior can result. In addition to the 'normal' solution in which all the balls are at rest for $t > t^*$, Lanford (1975) shows how to construct an anomalous solution in which all the balls are at rest for $t < t^*$ but for $t > t^*$ all but a finite number of them are in motion. This solution is obtained by taking the limit of standard solutions in which the spacing of the particles and the initial direction of the nth particle are arranged so that the nth particle just grazes the n-1th particle, sending it into a grazing collision with the n-2nd particle, etc. If the velocity of the nth particle increases rapidly enough as n increases, then the limiting solution as $n \to \infty$ contains within itself an analogue of the body of Fig. III.4 which appears from spatial infinity. If we plot successively the trajectories of the n-1st particle between the time when it is hit by n and the time it hits n-2, we get a zig-zag approximation to the trajectory β. Running this scenario backwards in time produces an infinite billiard ball analogue of the curve α of Fig. III.4 and an insult to historical determinism.

We have here a very curious situation. The billiard ball example conforms to Lucretius' vision of a world composed of nothing but atoms moving in a void. It also displays a non-deterministic spontaneity but not of the sort Lucretius thought necessary for free will, for not one of the billiard balls freely or spontaneously swerves in contravention to the laws of motion.

The self-exciting feature of Lanford's anomalous solution can be ruled out and determinism restored either by imposing population control and limiting the billiard game to a finite one or by setting boundary conditions limiting the behavior of the billiard balls at spatial infinity. Unless such limitations can be independently motivated we have yet another depressing example where determinism is achieved by fiat.

The need for the classical form of Laplacian determinism to appeal to boundary conditions at infinity arises not only in particle mechanics but in field theories as well, as is illustrated in the next section.

10. HEAT

The classical heat equation in one spatial dimension has a very simple appearance which belies the wealth of peculiarities it contains; it states:

$$(III.4) \quad \frac{\partial u}{\partial t} = \frac{\partial^2 u}{\partial x^2},$$

where for convenience the thermometric conductivity coefficient has been normalized to unity.

From the remark at the end of Sec. 5 it seems to follow that the heat equation cannot support any form of determinism, for it allows the scalar quantity u to vary in space and it is first order in time so that the appropriate initial data at $t = 0$ is $u(x, 0)$. However, the remark does not apply since the heat equation is not Galilean invariant. The intended physical interpretation of $u(x, t)$ is the temperature at time t and point x of some heat conducting medium, say, an iron bar. The rest frame of the medium is thus a preferred frame for describing thermal history. For the moment I will ignore the intended application and consider (III.4) as an abstract partial differential equation.

The abstract problem of Laplacian determinism is then to find a solution $u(x, t)$ of (III.4) satisfying initial conditions $u(x, 0) = \varphi(x)$, $-\infty < x < +\infty$ and to prove uniqueness of the solution. In this abstract form the problem is not well-posed, for there are null solutions to (III.4) which vanish at $t = 0$ but which are different from zero for $t > 0$. Since (III.4) is linear, such solutions may be added to any other solution to produce a new solution different from the original one but satisfying the same initial conditions at $t = 0$. Some null solutions are very smooth — indeed, C^∞ — so the breakdown in futuristic determinism is not due to the development of a singularity.[5]

Reflecting on the way heat is propagated according to (III.4) might make one despair of achieving any interesting uniqueness result for the initial value problem. (III.4) is a parabolic partial differential equation with characteristics coinciding with the planes of absolute simultaneity.[6] From this we deduce that heat is propagated infinitely fast so that influences coming from infinity would seem to be the norm. The deduction of infinite propagation velocity is supported by examining the fundamental source solution.

$$(\text{III.5}) \quad k(x, t) = \begin{cases} \exp(-x^2/4t) & \text{for } t > 0 \\ 0 & \text{for } t \leqslant 0 \end{cases}$$

Using the facts that for a solution u, $\int_a^b u(x, t)\, dx$ is taken to be the amount of heat contained in the medium at t between the points a and b, and that $\int_{-\infty}^{+\infty} k(x, t)\, dx = 1$, we interpret (III.5) as describing the temperature distribution of an initially cold bar into which a unit quantity of heat has been introduced at the origin. Since $k(x, t) > 0$ for any $|x| > 0$ and any $t > 0$, the heat is seen to spread instantaneously to the most remote parts of the medium.

However, the form of (III.5) also shows that the effects of the heat source are rapidly attenuated as one moves away from the spatial origin. This makes us hopeful that if the influences coming from infinity will not wreck uniqueness unless they are unboundedly strong. This hope is fulfilled with the aid of a formal condition limiting the growth of a solution u of (III.4) at infinity:

(B)   There are constants C and a such that

$|u(x, t)| < C \exp(ax^2)$

for all $-\infty < x < +\infty$ and all $t > 0$

Any solutions of (III.4) which satisfy (B) and which agree at $t = 0$ must agree for $t > 0$. Condition (B) can be weakened but not substantially; e.g., replacing x^2 by $|x|^{2+\varepsilon}$, $\varepsilon > 0$, does not secure uniqueness.

The boundary condition (B) might be promoted by the argument that if the temperature grows beyond all bounds the conditions of the problem are, physically speaking, undercut since our bar of iron will vaporize. But if determinism is going to break down it might as well go down with a bang, and complaining about the breakdown of the problem situation is just a way of bemoaning the demise of determinism. In any case, (B) can be violated without having the medium become as hot as Hades at spatial infinity; u might, to the contrary, become unboundedly negative.

This observation leads to another approach to establishing uniqueness. In view of the interpretation of u as temperature, we might want $u \geqslant 0$ everywhere and always. If our desire is fulfilled, then futuristic determinism is secured in this fashion: suppose that u_1 and u_2 solve (III.4), that $u_1, u_2 \geqslant 0$, and that $u_1(x, 0) = u_2(x, 0)$ for $-\infty < x < +\infty$; then $u_1(x, t) = u_2(x, t)$ for $t > 0$.[7]

For the intended application of (III.4) we may agree that $u \geqslant 0$ is a condition *sine qua non* for physical possibility. But what we would like is for this condition to result from a single initial stipulation and thereafter from the unfolding of determinism. That is, we would like

(H)   If u solves (III.4) and $u(x, 0) \geqslant 0$ for $-\infty < x < +\infty$, then $u(x, t) \geqslant 0$ for $t > 0$.

But (H) is false, as the alert reader will already have seen. For let u_1 and u_2 be any solutions of (III.4). By linearity, $\bar{u} = u_1 - u_2$ and $\bar{\bar{u}} = u_2 - u_1$ are also solutions. If u_1 and u_2 conform to the same initial conditions, then $\bar{u}(x, 0) = \bar{\bar{u}}(x, 0) = 0$, so by (H) $\bar{u}(x, t) \geqslant 0$ and $\bar{\bar{u}}(x, t) \geqslant 0$ for $t > 0$, implying that $\bar{u}_1 = \bar{\bar{u}}_2$, i.e., general uniqueness which we have seen does not hold. Thus in matters of heat once is not enough; the stipulation that $u \geqslant 0$ has to be repeated anew at each moment of time.

Finally, it may be well to note that the wringers through which heat puts determinism are not all peculiar to the heat equation. The type of field law appropriate for Newtonian space-time is a parabolic partial differential equation, the heat equation being only a special instance of the type. And for parabolic partial differential equations in general, uniqueness for the Laplacian initial value problem cannot be expected without the help of supplementary boundary conditions.

11. WALLING OUT

As an alternative to giving boundary conditions at infinity we could revert to the non-Laplacian wall strategy mentioned above. For the heat equation this would amount to specifying the function u both on the initial time slice S and on the boundary walls $W_1 \cup W_2$ (refer to Fig. III.3 again) and then trying to determine u within the interior region. This problem is well-posed under seemingly mild continuity assumptions, as follows from the maximum principle for parabolic partial differential equations. In the case of the heat equation, this principle asserts that if a solution u is uniformly continuous over the closed box of Fig. III.3, then u assumes its maximum value on the bottom or the side walls $S \cup W_1 \cup W_2$ of the box. To derive uniqueness, it suffices to take the case where $u = 0$ on $S \cup W_1 \cup W_2$; applying the maximum principle to both u and $-u$ gives $u \equiv 0$.

Instead of a portion of an infinite bar we can focus on a finite bar whose temperature at $t = 0$ is given and whose ends $x = 0$ and $x = 1$ are maintained by two stokers at prescribed temperatures over the interval from $t = 0$ to, say, $t = 1$. For this set up it is natural to require that the temperature $u(x, t)$ on the bar is continuous in x for any fixed t and continuous in t for any fixed point x of the bar. But as Hartman and Wintner (1950) note, the two-dimensional uniform continuity demanded by the maximum principle is not natural if we imagine that the stokers operate independently of one another and independently of the initial temperature distribution. But if two-dimensional uniform continuity is abandoned and what is required of a solution is ordinary continuity and the boundary conditions

$$u(x, 0^+) = \varphi(x), 0 < x < 1$$
$$u(0^+, t) = \psi(t), 0 < t < 1$$
$$u(1^-, t) = \kappa(t), 0 < t < 1$$

then the solution is not necessarily unique. Nor does the imposition of the one-sided boundedness condition $u \geqslant 0$ suffice for uniqueness for the modified problem. However, Hartman and Wintner show that ordinary continuity plus boundedness of u on the box do suffice for uniqueness. The latter condition is reasonable if we imagine that the stokers work with a finite fuel supply and stoke at a finite rate (conservationism and unionism in the service of determinism).

12. OLD HEAT

The problem of historical determinism for heat in an infinite bar has an
uninteresting dissolution. In the first place, the instantaneous temperature of the bar, regarded as the *final* temperature, cannot be arbitrarily
prescribed, for the assumption that heat has been diffusing according to
(III.4) for any length of time, no matter how short, forces $u(x, t)$ to
be analytic in x. Worse still, we know that without supplementary
boundedness conditions uniqueness cannot be expected; but boundedness conditions coupled with the assumption that the heat equation
holds for all past times tends to reduce the situation to an uninteresting
static one. Suppose that u satisfies (III.4) for all $t < 0$. If either $|u(x, t)|$
is uniformly bounded for $t < 0$ (i.e., there is a constant C such that
$|u(x, t)| < C$ for $t < 0$), or else $u(x, t) \geqslant 0$ for $t < 0$ and $u(x, 0)$ does
not grow too fast as $|x| \to \infty$, then $u =$ constant for $t < 0$ (Hirschman
(1952)). Interesting initial or final conditions can arise only if the
system is open, either to influences coming from infinity or to home
town stokers.

   Introducing stokers we can formulate a backwards final-boundary
value problem where the temperature distribution is known for the final
time S and for the ends of the bar W_3 and W_4 for earlier times, and the
temperature is sought for the interior R' of the bar at earlier times
(Fig. III.8). The maximum principle which was used to prove uniqueness for the forwards looking initial-boundary problem is asymmetrical
in time and does not yield uniqueness for the backwards looking
problem. (Recall: It asserts that for a closed box in the $x-t$ plane, the
temperature takes its maximum either on the *bottom* or the sides.)
Uniqueness is equivalent to the proposition that a solution which is 0
on $S \cup W_3 \cup W_4$ vanishes in the interior R'. Physically this would
mean that a finite bar whose ends are maintained at a zero temperature
cannot rid itself of all of its heat within a finite time.

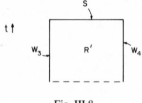

Fig. III.8

To investigate this possibility, assume that the temperature distribution at $t = 0$ is nice enough to be written as a Fourier series. For a bar extending from $x = 0$ to $x = 1$, this means that $u(x, 0) = \sum_{m=1}^{\infty} a_m \sin(m\pi x)$. The unique future solution is then $u(x, t) = \sum_{m=1}^{\infty} a_m \sin(m\pi x) \exp(-m^2 t)$. Uniqueness for the backwards final-boundary problem is then established by showing that for any finite $t > 0$, $u(x, t) = 0$ implies $u(x, 0) = 0$. Unfortunately, backwards uniqueness is of little help in practical cases of retrodiction since, as the form of the solution indicates, any error in the final data is exponentially expanded in trying to project into the past. Problems for prediction and retrodiction caused by instability will be discussed in Ch. IX.

It is of interest to note that if we append a non-linear term $f(u(x, t))$ to the classical heat equation equation, so that it now reads

$$\frac{\partial u}{\partial t} = \frac{\partial^2 u}{\partial x^2} - f(u),$$

then the solution of the modified equation can be driven to zero in a finite time, wrecking backward uniqueness.[8]

The heat equation can be used to make inferences about the past by using predictive models retrodictively. If we assume that at some time in the distant past the earth was in a molten state and neglect heat generated by chemical reaction, radioactivity, extra-terrestrial visitors, etc., then the heat equation can be solved forwards in time to give the present temperature distribution. In the simple models studied by Fourier and Kelvin, the temperature gradient near the earth's surface turns out to be proportional to the temperature of the molten material and inversely proportional to the square root of the period T since the molten state. In this way observations of current temperature gradients can be used to estimate the 'age' T of the earth. (Aside: Kelvin's estimate of 100—200 million years and later estimates by Tait which pushed the value down to 10—20 million years caused some consternation among geologists and the followers of Darwin. Of course, we now know that Kelvin's model contained a number of false assumptions.[9])

13. DON'T FENCE ME IN

We have seen that for determinism to succeed in Newtonian particle and field theories, either the erection of boundary walls or the imposition of boundary conditions at infinity is needed. For field theories these needs would disappear if space were compactified, eliminating spatial infinity. For particle theories, however, the situation is less clear. Roll up the Euclidean x–t plane along the x-axis to produce the cylindrical version of Newtonian space-time shown in Fig. III.9.

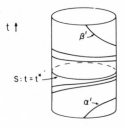

Fig. III.9

Even though there is no spatial infinity for particles to escape to or appear from, domains of dependence may still be trivial. The curves α' and β' result from α and β of Fig. III.4 when this figure is subjected to the rolling process. Since α' and β' have no end points, are everywhere time-like (i.e., are oblique to the planes of simultaneity and have finite velocities), but never meet S, $D(S) = D^+(S) = D^-(S) = S$. Similarly, for a three- or four-dimensional space-time, the initial-boundary value problem is threatened by particles which do higher dimensional analogues of the death spirals of those in Fig. III.9. Thus, in Fig. III.10, it seems that $D^+(S \cup W) = S \cup W$ and $D^-(S \cup W') = S \cup W'$.

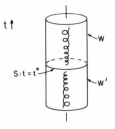

Fig. III.10

The point of erecting imaginary boundary walls was to prevent the unannounced invasion of influences from without. But if the processes illustrated in Fig. III.10 are live possibilities, then the invaders can invade from within. Examples of such fifth-columnist invaders cannot be constructed in Newtonian gravitational theory of point masses, at least not without collisions. A result of Sperling (1970) proves that if the solution ceases to exist after a finite time and there are no collisions, then the mutual particle distances cannot remain bounded; some of the particles must escape to infinity and in so doing will register on the walls. But perhaps fifth-columnists can be created with the help of binary collisions, as in the Mather-McGehee example, or by using a different kind of force function.

By now the determinist is becoming tired of having to fight a guerrilla war against the invaders who seek to overthrow a deterministic regime; but in Newtonian worlds there is no clear-cut path towards a once-and-for-all victory. For, to repeat, to incorporate into the space-time structure an unbreachable barrier to the invaders is to break Galilean invariance, and Galilean invariance is the Newtonian expression of the well-supported principle of the equivalence of inertial frames. Only a radical change in the structure of space-time can resolve this impasse in favor of the determinist. The special theory of relativity turns out to be an answer to the determinist prayers, as will be seen in the following chapter. But before leaving the classical domain, I want to discuss some other problems for determinism, one of which does and the other of which does not derive from very fast particles or waves.

14. CLASSICAL ELECTROMAGNETISM

The source free Maxwell equations in empty space read

$$(\text{III.6}) \quad \nabla \times \mathbf{E} = -\frac{1}{c}\frac{\partial \mathbf{B}}{\partial t} \qquad \nabla \times \mathbf{B} = \frac{1}{c}\frac{\partial \mathbf{E}}{\partial t}$$

$$\nabla \cdot \mathbf{B} = 0 \qquad\qquad \nabla \cdot \mathbf{E} = 0$$

Since these equations are not Galilean invariant they require the support of a special frame of reference. In the 19th century this frame was taken to be the rest frame of a ponderable medium, the luminiferous aether, which was thought to be a necessary substratum for electromagnetic waves. However, in keeping with the dematerialization of the aether which took place at the turn of the century, I will construe the aether frame to be a special inertial frame, absolute space, which is unoccupied by a material substratum.[10]

It then follows from (III.6) that electromagnetic waves are propagated with a speed c relative to absolute space and a speed $c \pm v$ in a
frame moving relative to absolute space with a speed v. This puts the
theory into conflict with actual observational results, but let us imagine
that Nature has spoken against Galilean invariance and then ask
whether the theory provides us with an example of determinism. It
does. If the values of **E** and **B**, subject to the instantaneous constraints
$\nabla \cdot \mathbf{B} = \nabla \cdot \mathbf{E} = 0$, are specified at one time, then the top two of the
Maxwell equations determine the future values, guaranteeing in the
process that the constraint equations continue to hold.

This success for determinism becomes tainted when we attempt to
add sources. Formally, the second of the top two Maxwell equations is
modified by adding the current density to the right hand side and the
second of the bottom two is modified by adding the charge density to
the right hand side. The theory is completed by adding the Lorentz
force law governing the motion of charges. The resulting formalism
admits a well-posed initial value problem as long as the charges move
with subluminal velocities, but there is nothing in the formalism as
stated to prevent the presence of charged tachyons. With tachyonic
sources it remains to be seen whether the system admits a coherent
initial value problem and, if so, whether there are solutions in which the
tachyons accelerate themselves off the space-time manifold. If in either
of these ways classical tachyons should undermine Laplacian determinism we could consider modifying the laws of classical electromagnetism so as to prevent a charged particle from becoming a
classical tachyon by accelerating itself from a sub to a superluminal
velocity. But this already takes us part way towards relativity theory.
And it leaves unexplained why classical charged tachyons couldn't have
existed from time immemorial in a superluminal state of motion.

15. SHOCK(ING) WAVES

We made repeated use of two characteristics of the classical heat
equation: linearity and infinitely fast propagation of heat. Another
interesting feature is that the heat equation has a soothing effect on
temperature: whatever roughness exists in the initial temperature distribution is smoothed out in ever so short a time, for solutions $u(x, t)$
become analytic in x (though not in t) for $t > 0$.

Hyperbolic partial differential equations imply finite signal velocities. But non-linear versions of these equations may not have the soothing effect of the heat equation; indeed, solutions may shed whatever smoothness exists in the initial data and become non-differentiable or even discontinuous, thus ceasing to exist as ordinary solutions.

A very simple example studied intensively by mathematicians is the first order equation.

(III.7) $\dfrac{\partial u}{\partial t} + \dfrac{\partial f}{\partial x} = 0$ (where $f = f(u(x, t))$).

Setting $f' = df/du$, (III.6) can be rewritten as

(III.8) $\dfrac{\partial u}{\partial t} + f' \dfrac{\partial u}{\partial x} = 0.$

It follows that a solution u is constant along the characteristic curve $x(t)$ which has velocity

(III.9) $dx(t)/dt = f'(u(x, t))$

To get a solution corresponding to initial data $u(x, 0) \equiv u_0(x)$, we just propagate the initial data along the characteristics thus: $u(x, t) = u_0(x - tf')$. In the linear version of (III.8), $f' = $ constant and the characteristics are independent of the particular solution. Also the characteristics radiating from the line $t = 0$ simply cover the upper half of the $x-t$ plane, and thus the initial data at time 0 can be propagated forward in time to give a solution for all $t > 0$. But suppose that (III.8) is genuinely non-linear with, say, $f'' > 0$. If the initial data are chosen so that $u_0(x_1) > u_0(x_2)$, $x_1 < x_2$, then the characteristic radiating from $(x_1, 0)$ has a greater velocity than the one from $(x_2, 0)$. So the two must intersect at some point (x^*, t^*) with $t^* > 0$, with the result that u takes on two different values at the same point. Solutions in the ordinary sense may fail to exist after a finite time.

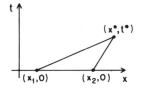

Fig. III.11

Anarchists will be happy to conclude that the law of motion breaks down and chaos reigns. Those more disposed towards law and order will seek a generalized sense of 'solution' on which u can be said to solve (III.8) even though it is not differentiable or even continuous. The mathematical theory of distributions is tailor made for this situation. u is said to be a *weak solution* of (III.7) in the sense of distributions just in case

$$(\text{III.10}) \quad \iint \left[\frac{\partial \phi}{\partial t} u + \frac{\partial \phi}{\partial x} f \right] dx\, dt = 0$$

for every test function $\phi(x, t)$ which is C^∞ and which vanishes outside of some compact region of the x–t plane.

Weak solutions overcome the existence problem, but at the expense of uniqueness since more than one weak solution can correspond to the same initial data.[11] The committed determinist will be convinced that the space of all weak solutions is too large, that it extends beyond the bounds of real physical possibility, and that uniqueness will be restored when the unphysical solutions are cut out. But lest he be accused of chicanery in cutting out solutions, the determinist must allow his hand to be guided by independent considerations as to what is and is not physically possible. Just such a guide comes from experience with shock waves, which provide the physical motivation for studying weak solutions in the first place. As a piston compresses a cylinder of gas it creates a wave which travels through the gas with the speed of sound s. But as the gas is compressed, s increases so that the later waves move faster than the earlier ones. In some conditions, the later waves overtake the earlier ones and in such a way that the resulting waveform develops a shock discontinuity where the velocity gradient blows up.[12] If u in our equation (III.7) is interpreted as the velocity of the gas, then it provides a simple mathematical model for the formation of shocks. The velocity gradient is

$$(\text{III.11}) \quad \frac{\partial u}{\partial x} = \frac{u_0'}{1 + u_0' t f''}$$

In keeping with the above assumptions, we set $f'' > 0$ and $u' < 0$ and find that a gradient catastrophe occurs at the positive time $t = -1/u'f''$.

Suppose then that we assume that the only way ordinary solutions degenerate into weak ones is through the formation of shocks, which we will idealize in the following way: there is a smooth curve $y = x(t)$ across which the solution u may be discontinuous but on either side of which it is smooth. The determinist will then want to show that corresponding to any initial value problem there is a weak solution of this form. This can be done, but alas it still may not be unique. Further surgery on the class of weak solutions is required. If we think of the formation of shocks as an irreversible process, it is natural to require that matter which crosses the shock show an increase in entropy. Analytically, this amounts to requiring that the velocity dy/dt of the shock is less than the characteristic velocity $f'(u_l)$ on the left but greater than that $f'(u_r)$ on the right; or equivalently, each point on the shock properly reflects the initial data by being connectible to the initial data surface by a characteristic. With these restrictions in place, uniqueness of weak solutions can be proved, if, as we assumed, $f'' > 0$. If f' is not an increasing function of u a more complicated form of the entropy condition is needed.[13]

For the determinist, the lesson to be drawn is clear. The apparent problem with determinism was a welcome opportunity to investigate in detail the physics of the situation and to show that when that is done determinism works its way in a more subtle and wondrous form than we could have otherwise imagined. The skeptic will complain that the determinist should have been able to say in advance what all the constraints were and should not have been allowed to cut the cloth of physical possibility to suit the needs of determinism.

16. VISCOUS FLUIDS

As a final example, I mention the Navier-Stokes equation, which is the classical equation of motion for a viscous fluid. For appropriate initial data at $t = 0$, a regular solution is known to exist at least for a finite interval $[0, t^*)$, and when it exists it is unique. A weak solution exists for all future time, and in the case of two-dimensional motion is unique. But global uniqueness for weak solutions in real three-dimensional space remains an open question.[14]

In the 1930's Leray (1934) conjectured global uniqueness does not hold for all initial data and that the breakdown of uniqueness (in weak solutions) is associated with the development of turbulence in the fluid. More recently opinion has swung away from Leray's point of view towards determinism and towards an alternative explanation of turbulence, advocated by Ruelle (1981), in terms of "strange attractors." If the evolution of the fluid is indeed deterministic, then its possible motions can be described as a flow on a phase space, each point of which represents a possible instantaneous state of the system. An attractor A is a point or more generally a compact region of the phase space such that the phase orbit uniquely determined by any point sufficiently near to A converges upon A. Conservative dynamical systems (e.g., those described by Hamiltonian mechanics) where the phase flow preserves volume in phase space cannot have attractors. But dissipative systems, such as viscous fluids, where the internal friction dissipates mechanical energy, generally do have attractors. An attractor A is strange if, roughly, the phase orbits determined by points near A are unstable. More will be said about these issues in Ch. IX. The only point I wish to convey here is that determinism in the classical description of this most earthy of processes (tea sloshing in a cup, whirlpools in rivers, etc.) is very much a live issue.

17. CONCLUSION

Several important morals can be drawn from our discussion of determinism in classical worlds. The overarching moral concerns the importance of the status and structure of space-time. On a Newtonian substantivalist conception of space-time, Laplacian determinism is not a free-standing doctrine but requires sufficiently strong space-time scaffolding to support it. A Leibnizian non-substantivalist conception of space-time may avoid the need for some of the scaffolding, but the Leibnizian alternative has never been worked out in sufficient detail to permit judgments to be made with any confidence.[15]

Newtonian space-time, whose structure is rich enough to support the possibility of Laplacian determinism, nevertheless proves to be a none too friendly environment. The principal irritant derives from the possibility of arbitrarily fast causal signals, threatening to trivialize domains of dependence. It is not surprising, therefore, to find non-uniqueness for the initial value problem for some of the most fundamental equations of motion of classical physics, both for cases of discrete particles (ordinary differential equations) and for continuous media or fields (partial differential equations). Whether such non-uniqueness entails the falsity of determinism is a difficult and delicate question, turning in large part on the status of supplementary conditions that might be imposed on the problem. We encountered a variety of such cases, ranging from those where the supplementary conditions needed to restore uniqueness are both physically well-motivated and nonquestion-begging to others where the supplementary conditions amount to little more than a hypocritical refusal to consider the possibility of unpleasant surprises. Individual attitudes on the classification of cases is naturally influenced by one's predispositions towards determinism. Such a circularity is not unexpected; nor is it entirely unwelcome since it provides a means by which determinism can be used to probe issues about physical possibility and necessity.

Though they are perhaps obvious, there are two other points worth emphasizing. First, the trials and tribulations determinism is forced to undergo in classical physics are purely ontological. None of the ones I have described above derive from epistemological considerations, such as the ability of observers, embodied or disembodied, smart or dumb, to access and process information about the universe. Second, despite the residual and irremediable vagueness in the ontological doctrine of determinism, the threats discussed above are sharp enough to be recognizably threats. And I would add, the issues are not sharpened by yielding to the current philosophical fashion of formalization. If philosophers had spent less time fiddling with axioms, subscripts, n-tuples, and the like, and more time on physics, they would no doubt have produced a better assessment of classical determinism than I have managed.

Whatever the outcome on the substantive issues, it is clear that the long-standing confident pronouncements about classical determinism have been premature. It wasn't until quite recently that hard mathematical results on existence and uniqueness were obtained, and important questions remain open. Classical determinism is not the mummified relic that philosophical literature portrays it to be, but a living and breathing creature capable of generating surprising twists and turns.

NOTES

[1] Following the discussion in Sec. 2 below on the connection between space-time symmetries and symmetries of laws, the condition for Laplacian determinism can be formulated as follows. For any physically possible $\langle M, \{G_a\}, \{P_\beta\} \rangle$ and $\langle M, \{G_a\}, \{P'_\beta\} \rangle$

and any diffeomorphism $d: M \to M$ (onto) and any plane of absolute simultaneity $S \subset M$, if $d(S) = S$, $d^*G_a = G_a$, and $d^*P_\beta(p) = P'_\beta(p)$ for all $p \in S$, then $d^*P_\beta = P'_\beta$ everywhere. (d^*O denotes the dragging along of the object O by d.) For reasons which will emerge below, Leibniz would have wanted to weaken the requirement of determinism as follows: for any physically possible $\langle M, \{G_a\}, \{P_\beta\} \rangle$ and $\langle M, \{G_a\}, \{P'_\beta\} \rangle$ and any diffeomorphism $d: M \to M$ (onto) and any plane of simultaneity $S \subset M$, if $d(S) = S$, $d^*G_a = G_a$, and $d^*P_\beta(p) = P'_\beta(p)$ for all $p \in S$, then there is a diffeomorphism $d': M \to M$ (onto) such that $d'^*G_a = G_a$ and $d'^*P_\beta = P'_\beta$ everywhere.

[2] See my (1979) paper for some ideas on how this trick might be accomplished.

[3] This example is taken from Pollard (1966). The somewhat dated though still valuable classic reference on celestial mechanics is Wintner (1947).

[4] This point is brought out in Geroch (1977).

[5] Widder's book (1975) is the definitive survey on what is known about the heat equation.

[6] See John (1982) for definitions and examples of the concepts used in analyzing partial differential equations.

[7] This result was first obtained in 1944 by Widder; see Widder (1975).

[8] See Payne (1975). This monograph discusses various examples of partial differential equations where uniqueness and/or stability fail.

[9] See sec. 28 ("The Age of the Earth") of Carslaw (1921).

[10] We do not have a fully specified theory of electromagnetism until we have said what kind of geometric objects $\mathbf{E}$ and $\mathbf{B}$ are or, equivalently how $\mathbf{E}$ and $\mathbf{B}$ transform under Galilean transformations. It is known that there are two possibilities, both of which have unattractive features; see Earman and Glymour (1982).

[11] See Lax (1973), from whom the present discussion is taken.

[12] The classic reference to the physics of shock waves is Courant and Friedrichs (1976).

[13] For details, see again Lax (1973).

[14] See Temam (1983) for a survey of what is currently known about existence and uniqueness for the Navier-Stokes equation.

[15] But note for future reference that a non-substantivalist conception of space-time is needed in the context of general relativity theory if determinism is to stand a fighting chance of being true; see Ch. X below and Earman and Norton (1986).

SUGGESTED READINGS FOR CHAPTER III

None. The truth about determinism in classical physics, as I have tried to indicate in this chapter, is both fascinating and complex. Unfortunately, it lies buried in technical treatises and research papers in mathematical physics. I have no magic set of instructions for extracting it; if you are interested, you just have to start digging, following the leads provided above.

CHAPTER IV

DETERMINISM IN SPECIAL RELATIVISTIC PHYSICS

> Henceforth space by itself, and time by itself, are doomed to fade away into mere shadows, and only a kind of union of the two will preserve an independent reality.
>
> (Hermann Minkowski, "Space and Time")

The revolution which issued in the special theory of relativity (STR) was not prompted by self-conscious reflections on the problems and prospects of determinism in Newtonian physics. And yet there is a remarkable overlap in the considerations which might have convinced the dyed-in-the-wool determinist that something was badly amiss in the classical conception of space and time and the considerations which led Einstein to the relativistic conception. The determinist who does not want to fight a never ending guerrilla war against invaders from infinity needs a finite bound V_{max} on the speed of causal propagation. But if there is such a V_{max}, then the operational significance of Newtonian absolute simultaneity is called into doubt. The operational meaning of clock synchronization is the subject of Section 1 of Einstein's 1905 paper "On the Electrodynamics of Moving Bodies" which is generally acknowledged to have laid the foundations of STR. Again, a finite invariant velocity requires the introduction of absolute space, or its equivalent, and the breaking of Galilean invariance. But Galilean invariance is the classical expression of the lawlike equivalence of all inertial frames, a principle supported by the great weight of experience. Einstein also struggled with a version of this conflict in electrodynamics. Both theoretical reasoning and experiment convinced him of two things; first, that the (special) principle of relativity for mechanics extends to electromagnetism in that "the same laws of electrodynamics and optics will be valid for all frames of reference for which the equations of mechanics hold good"; and second, that "light is always propagated in empty space with a definite velocity c." One of the stated aims of Einstein's 1905 paper is to reconcile these "apparently irreconcilable" postulates.[1] How Einstein actually arrived at the STR and whether the order of discovery paralleled the order of presentation in the 1905 paper is a matter of controversy among historians of science.[2] But such controversies need not detain us since we are interested in reading the message of the completed theory for determinism.

In the popular imagination Einstein's contribution to the discussion of causality consisted of an heroic but fruitless tilting with the quantum mechanicians over the question of whether God plays dice with the universe. In fact, Einstein's contributions to determinism were profound — both profoundly positive and profoundly negative. Before his special theory of relativity unproblematic examples of determinism were hard to come by; after his general theory they are equally hard to find. This chapter details the positive contribution while Ch. X explores the negative contribution.

1. SPECIAL RELATIVISTIC WORLDS

Two of the three key assumptions characterizing classical worlds remain intact for the special relativistic case: all the physically possible special relativistic worlds are assumed to have a common fixed space-time background, and the physical contents of each such world are specified by giving the values of space-time magnitudes. However, as the quotation from Minkowski indicates, Newtonian absolute simultaneity vanishes.

The layman is apt to start with the prejudice that relativistic space-time is a strange and complicated affair. Strange it may seem at first, but complicated it is not. If the language of modern differential geometry is used to compare Newtonian and relativistic space-times, it is the former which comes off as complicated. To characterize the structure of Newtonian space-time requires several distinct geometrical objects and a number of mathematical conditions to assure the proper meshing of these objects.[3] But, as Minkowski first showed, all of the structure of special relativistic space-time (hereafter, Minkowski space-time) flows from a single object, the Minkowski metric.

Minkowski space-time consists of a space-time manifold M, usually assumed to be the standard $\mathbb{R}^4$ as in the classical case, and the Minkowski metric η. Technically (and pedantically), η is a symmetric, non-singular tensor field of type $(0, 2)$ and signature $(+++-)$. 'Type $(0, 2)$' means that η maps pairs of tangent vectors U, V on M into real numbers $\eta(U, V)$. Symmetric means that $\eta(U, V) = \eta(V, U)$. The indefinite signature means that the tangent space at each point $x \in M$

possesses a cone structure which divides the space into three disjoint regions as follows: V is *timelike* (respectively, *spacelike, null* or *lightlike*) according as $\eta(V, V) < 0$ (respectively, $\eta(V, V) > 0$, $\eta(V, V) = 0$). The timelike portion of the cone has two separate lobes, and Fig. IV.1 follows the usual convention of choosing the upper one to be the future lobe and the bottom one the past lobe. How past and future are distinguished by physics itself is part of the problem of the direction of time, a problem we will bump into from time to time.

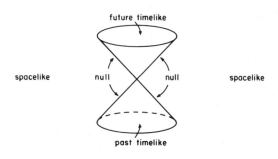

Fig. IV.1

Part of what makes the STR special is the further assumption that η is flat or pseudo-Euclidean. On $M = \mathbb{R}^4$ this means that it is possible to choose a global coordinate system (x^a, t), $\alpha = 1, 2, 3$ in which the components η_{ij} $(i, j = 1, 2, 3, 4)$ of η have the form $\eta_{ij} = \text{diag}(+1, +1, +1, -1)$. Such a system is said to be inertial. Suppressing two spatial dimensions, the relation of two such inertial systems (x, t) and (x', t') whose spatial origins coincide at $t = 0$ is found to be given by

$$(L) \qquad x' = \frac{x + vt}{\sqrt{1 - v^2}}, \qquad t' = \frac{t + vx}{\sqrt{1 - v^2}}$$

For convenience, the velocity of light c has been normalized to unity. Putting c back into (L), one has the more familiar form of the Lorentz transformations:

$$(L') \qquad x' = \frac{x + vt}{\sqrt{1 - v^2/c^2}}, \qquad t' = \frac{t + vx/c^2}{\sqrt{1 - v^2/c^2}}$$

The invariance of the velocity of light is built into the geometry *ab initio*, but if you prefer coordinate language, the constancy of c can be verified by plugging into the relativistic velocity addition formula $u' = u + v/(1 + uv/c^2)$, which follows from (L').

A useful heuristic for thinking about the relation between Minkowski space-time and Newtonian space-time is to view the latter as the result of the former in the limit as c "goes to infinity." Intuitively, the light cones collapse into planes of absolute Newtonian simultaneity and algebraically (L') goes over into the Galilean transformations. (To make this heuristic precise, one could, for example, choose a timelike inertial direction field on Minkowski space-time and then collapse the light cone at each point symmetrically about the chosen direction.[4] The limiting process recovers more than we may have wanted — not only absolute simultaneity but absolute space as well.)

Our previous discussion of absolute time coupled with this heuristic suggests that in relativistic worlds c plays the role of the maximum signal velocity. This will be taken for granted until Sec. 9 below.

2. DOMAINS OF DEPENDENCE

Domains of dependence in relativistic space-times are defined as before, except that a causal curve is taken to be one which lies inside or on the light cone rather than one which lies oblique to a plane of absolute simultaneity. But to be more specific, a space-time curve is a differentiable map $\sigma: I \to M$ from some interval $I \subset \mathbb{R}$ into space-time M. (Sometimes 'curve' will also be used to denote the image set of the map.) σ is *timelike* (respectively, *null, causal*) according as $\dot{\sigma}(\lambda)$, $\lambda \in I$, is timelike (respectively, null, null or timelike). Such a σ is *future* (respectively, *past*) *directed* according as $\dot{\sigma}(\lambda)$, $\lambda \in I$, is future (respectively, past) pointing. For $R \subset M$ the future domain of dependence $D^+(R)$ of R is defined as the collection of all points $x \in M$ such that every future directed causal curve which passes through x and which has no past end point meets R. The past domain of dependence $D^-(R)$ of R is defined analogously.[5]

Pictured in Fig. IV.2 are some typical domains of dependence for Minkowski space-time. Since these domains are non-trivial, Laplacian determinism seems at least possible.

3. THE RELATIVISTIC FORMULATION OF
LAPLACIAN DETERMINISM

Though much more friendly towards determinism than its Newtonian

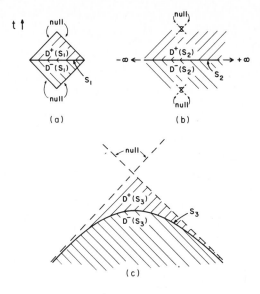

Fig. IV.2

counterpart, the special relativistic environment still contains pitfalls. Any spacelike hypersurface can be thought of as a surface of simultaneity, and any global spacelike hypersurface (i.e., one which has no edges) can be thought of as a time slice, the state on which is the 'world at a given time'. This might lead us to try to ape the classical definition as follows: world $W \in \mathcal{W}$ is *futuristically* (respectively, *historically*) *deterministic* just in case for any $W' \in \mathcal{W}$, if W and W' agree on any time slice, then they agree everywhere. In this form the doctrine of Laplacian determinism is doomed to failure, as the reader who has bothered to look at Fig. IV.2 will have already deduced. The spacelike surface S_3 in Fig. IV.2(c) is a time slice, but its domain of dependence includes nothing outside of the past lobe of the pictured null cone. To stand a chance of being true, the doctrine must be more circumspect in its choice of time slices.

Any time slice S in Minkowski space-time divides all events into three disjoint classes: those lying on S itself; those lying to the future $F(S)$ of S and those lying to the past $P(S)$ of S. Call S a *future* (respectively, *past*) *Cauchy surface* just in case $F(S) \subset D^+(S)$ (respectively, $P(S) \subset D^-(S)$). S is a Cauchy surface simpliciter just in case it is both past and future Cauchy, or what comes to the same thing, $D(S) =$

$D^+(S) \cup D^-(S)$ is the entirety of Minkowski space-time. We can now avoid the above pitfall by restricting the definition of global Laplacian determinism to time slices which are Cauchy.

There is a slight embarrassment in imposing this restriction because the statement that S is Cauchy is not a statement about S alone or about a finite neighborhood of S but rather a statement about the entire space-time. To see this, note that if a single point is removed from the future side of S (see Fig. IV.3), then S is no longer a future Cauchy surface. No point in the shaded region belongs to $D^+(S)$ and, in fact, this punctured space-time contains no Cauchy surfaces at all. The embarrassment can be overcome by reiterating the initial stipulation that in special relativistic worlds the space-time is given once and for all as Minkowski space-time. Cauchy surfaces thus always exist and, further, can always be recognized by intrinsic features. If, for example, the space metric induced on S by the Minkowski space-time metric is Euclidean and S has no edges, then S is Cauchy for Minkowski space-time.

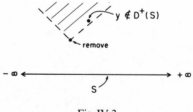

Fig. IV.3

In a general relativistic setting Cauchy surfaces do not always exist, and when they do they cannot always be recognized by their local characteristics and the embarrassment returns. But that is a matter to be discussed later; for now, let us enjoy determinism while we can.

4. LAPLACIAN DETERMINISM — AT LAST!

We have non-trivial domains of dependence. We have Cauchy surfaces. And we have more than enough space-time structure to avoid the pitfall of Leibnizian determinism, for any isomorphism of Minkowski space-time (i.e. any diffeomorphism of $\mathbb{R}^4$ onto itself which preserves η) that leaves fixed pointwise a Cauchy surface (say, $t =$ constant for some inertial coordinate time t) must be the identity map. Laplacian determinism thus has a secure launching pad in Minkowski space-time. But we still need concrete examples of how in real physical processes determinism can soar to the lofty heights of James' vision.

Perhaps the simplest of all field equations that are Lorentz invariant and have physically important applications is the homogeneous wave equation. In one spatial dimension it reads

$$\text{(IV.1)} \quad \frac{\partial^2 u}{\partial t^2} - \frac{\partial^2 u}{\partial x^2} = 0$$

It propagates influences with velocity c ($\equiv 1$) as can be deduced from its characteristic equation ($x \pm t = 0$) or from the form of the solution to the initial value problem, which we now formulate.

At $t = 0$ we prescribe the initial data

$$\text{(I)} \quad u(x, 0) = f(x)$$
$$\frac{\partial u(x, 0)}{\partial x} = g(x) \quad -\infty < x < +\infty$$

Assuming that f is C^2 and g is C^1, there is a unique solution for $t > 0$, and it can be written in explicit form:

$$\text{(IV.2)} \quad u(x, t) = \tfrac{1}{2}(f(x+t) + f(x-t)) + \int_{x-t}^{x+t} g(\xi)\, d\xi$$

No boundary conditions at infinity are imposed. No assumption that the solution is C^2 is imposed; that follows from the problem set up. No handwaving, no extra props, no pious prayers are needed. Given only the law and the initial data of enough smoothness to assure that the law holds in the ordinary sense at $t = 0$ we are guaranteed of a unique solution for all future times. (A similar result holds for the past direction of time.) No equivocation or shadow of turning in the past or future is possible. Here is Laplacian determinism triumphant — at last!

5. HIGHER DIMENSIONS AND WEAK SOLUTIONS; HUYGENS' PRINCIPLE

The triumphant celebration above was for the flat-land of one spatial dimension. In real three-dimensional space things are more complicated because the wave equation can focus roughness in the initial data from different regions of space into a smaller region, causing a greater roughness. But since only one degree of differentiability can be lost by focusing, we could make sure of ordinary solutions by requiring that the initial data functions f and g (now considered as functions on $\mathbb{R}^3$) are respectively C^3 and C^2. If the physics of the situation does not conform to this requirement then a different approach is needed. For finite systems a satisfactory approach exists.

Let us say that a solution u represents a finite system at time t if (a) $u(x, t)$ and its first derivatives vanish except for x in some compact region of space, and (b) at t the energy of the system

$$(IV.3) \quad E_u(t) \equiv \iiint \left[\left(\frac{\partial u}{\partial t} \right)^2 + \sum_{a=1}^{3} \left(\frac{\partial u}{\partial x^a} \right)^2 \right] dx$$

is finite. It is easy to see that if the system is finite at $t = 0$, it is finite at any later time. For by finite speed of propagation, if (a) holds at $t = 0$, it holds at any $t > 0$. Then by differentiating $E_u(t)$ with respect to t and using (IV.1), it follows not only that energy remains finite, but it also remains constant — finite systems are conservative.

The weak solutions of the wave equation for finite systems can be obtained as limit points of the space of ordinary solutions where convergence is measured in the 'energy norm' (the square root of the energy). These limiting solutions are indeed solutions in the sense of distributions and they restrict to initial value hypersurfaces so that the Laplacian initial value problem is meaningful. For initial value functions f and g such that $\partial f / \partial x^a$ ($a = 1, 2, 3$) and g are square integrable, uniqueness follows from conservation of energy. (If u_1 and u_2 are solutions with the same initial data, then $\bar{u} = u_1 - u_2$ is also a solution with $E_{\bar{u}}(0) = 0$. By energy conservation, $E_{\bar{u}}(t) = 0$, $t > 0$. Since the integrand of energy is non-negative, $\bar{u} = \text{constant} = 0$.)

The condition (a) for finite systems can be relaxed for quasi-finite systems where u falls off sufficiently rapidly as $|x| \to \infty$. For systems which are not even quasi-finite, local energy conservation can be maintained by erecting walls and taking into account the energy flux through the walls. However, because of the finite speed of propagation of the field, a walls strategy is not needed to secure local determinism.

Part of the content of Huygens' Principle can be captured as a statement about determinism. Choose any event p in Minkowski space-time and any Cauchy surface S lying to the past of p. Then the intersection $C^-(p) \cap S$ is a compact subset of S. (Here $C^-(p)$ denotes the *causal past* of p, i.e., the set of all space-time points q such that there is a (possibly trivial) future directed causal curve from q to p.) Huygens' Principle then asserts that the state on an arbitrarily small neighborhood of this intersection uniquely determines the state at p. Effectively, disturbances propagate with *exactly* the speed of light, making it possible to construct sharp signals which die out completely at a spatial location within a finite time after the source of the disturbance has been 'switched off'. The wave equation in Minkowski space-time satisfies Huygens' Principle if and only if the number of spatial dimensions is odd (John (1983)). Hyperbolic partial differential equations in general do not exhibit Huygens' Principle, for they typically permit disturbances to propagate with a velocity less than that of light and the disturbance can 'ring on' indefinitely after the source has been switched off. In four-dimensional general relativistic space-times the wave equation exhibits Huygens' Principle only under very special conditions; in empty space solutions to Einstein's field equations without cosmological constant, i.e., the Ricci tensor vanishes (see Ch. X below), the conditions are that the space-time is flat or else is a plane wave solution to Einstein's field equations (McLenaghan (1969)).

6. DOMAINS OF PREDICTION

Questions about scientific predictability are often posed in terms of disembodied spirits whose intelligence may range over the entire spatial extent of the universe (recall Laplace's demon) or in terms of embodied observers who are given information about the past and present state of the world. But this approach leads to a never-never land form of prediction that is unavailable to actual observers who are localized and embodied and who are not 'given' any free gifts of information but must ferret it out for themselves.

Towards a more realistic sense of scientific predictability, let us try to define the notion of the *domain of prediction DP(R)* of a space-time region R. Two constraints seem necessary. First, if events at a space-time point x are to be predictable form R then from the perspective of R they should not have already occurred; and second, if the events at x are to be scientifically predictable from R then they must be in principle determinable from the state at points from which it is physically possible for R to receive information. To capture these constraints, define the *causal past* $C^-(R)$ of R to be $\cup_{p \in R} C^-(p)$. Then the first constraint can be stated as the condition that if $x \in DP(R)$, then $x \notin C^-(R)$. And the second constraint demands that if $x \in DP(R)$ then $x \in D^+(C^-(R))$.

If these two constraints are sufficient as well as necessary, then we can set about computing domains of prediction for various choices of R. If, for example, R is a Cauchy surface of Minkowski space-time, then as expected, $DP(R)$ consists of all the points to the future of S. To apply this apparatus to the question at issue, we need to decide on an appropriate R for embodied observers.

Consider first an idealized dimensionless observer whose world line ν is pictured in Fig. IV.4(a). At any moment in his existence, say, p,

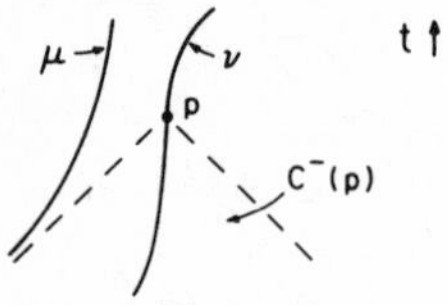

Fig. IV.4(a) 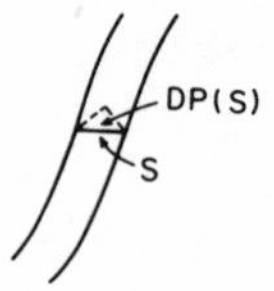   Fig. IV.4(b)

$C^-(p)$ contains all those events about which the observer can gain knowledge by direct observational means. The appropriate R for that moment is thus p. But for any point p of Minkowski space-time, $D^+(C^-(p)) = C^-(p)$, and so $DP(p) = \phi$. The observer can predict exactly nothing. The more well-fed observer represented in Fig. IV.4(b) by a world tube will have a non-empty domain of prediction associated with his spatial extent at any moment. A moment in this fellow's life is given by a spacelike slice S through his world tube, and $DP(S) \neq \phi$. But note that $DP(S)$ lies entirely within his own corpulence so that, at best, he can only predict the rumblings of his own innards. And even this self-prediction is an unreachable ideal, for by the time the information about the state of his body is transmitted along whatever serves as his nervous system to a central processing location, the events to be predicted will be past history to him.

We have arrived at a curious situation. The structure of Minkowski space-time makes futuristic Laplacian determinism possible by shielding against the invasion of unsettling influences coming in from spatial infinity. But the very same structure seems to make it impossible for embodied localized observers to cash determinism in for prediction. For Popper there is no paradox here since he takes the above considerations to prove that determinism cannot be true in special relativistic worlds.[6] I invite the reader to conclude that, to the contrary, we have yet another reason to reject the equation of determinism with predictability.

For the determinist who sees the cash value of determinism in prediction, the modified two-dimensional Minkowski space-time pictured in Fig. IV.5 is a dream come true. The causal past of any point p wraps around the universe so that any causal curve without past end point (e.g. γ) must enter $C^-(p)$. Thus, $DP(p)$ includes every point in the complement of $C^-(p)$.[7] While examples of this sort are mathematically intriguing, they are of little use to the observers who inhabit not dream worlds but the postulated standard Minkowski worlds.

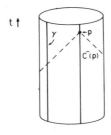

Fig. IV.5

Another way to restore predictability and boost determinism at the same time would be to (i) ban source-free photons and (ii) require that all photon sources as well as all nonzero rest mass particles do not, as one goes backwards in time, asymptotically approach the velocity of light as does the particle whose world line is labeled μ in Fig. IV.4(a). Computing domains of prediction using causal curves conforming to (i) and (ii) has the effect that in Minkowski space-time $D^+(C^-(p))$ is not only not empty but includes the complement of $C^-(p)$. One can imagine a plausible scenario backing (i), but enforcing (ii) is going to be even trickier than enforcing the ban on the Newtonian invaders from infinity.

The only alternative is to admit that even with determinism, prediction in special relativistic worlds is a chancy affair. Using determinism, we can form, with complete confidence, hypothetical predictions, where the hypotheses concern the state of the infinite universe or else the influences that will or won't penetrate the boundaries of a finite system. Observation combined with inductive reasoning may recommend one hypothesis over all others, but rarely, if ever, does the combination yield a confidence that approaches the certainty with which Laplace's demon went about its prediction tasks.

7. PARTICLE MOTION: RETARDED AND ADVANCED ACTION-AT-A-DISTANCE

A fair comparison of Newtonian and special relativistic worlds would seem to demand that we examine the relativistic analogues of the classical heat equation and the Newtonian equations of particle mechanics. Unfortunately, there is no generally accepted relativistic phenomenological heat equation. But it is assumed that any acceptable candidate must be a hyperbolic partial differential equation, implying finite speed of propagation for heat and making possible a well-posed Laplacian initial value problem without need to impose boundary conditions at infinity. The discussion of relativistic particle mechanics is clouded by the widely held opinion that the spirit, if not the letter of STR, implies that particles cannot act at-a-distance but must transmit their influences by means of a field mechanism. Despite this opinion, the physics literature contains a large and ever growing list of self-consistent and Lorentz invariant theories whose only state variables are particle variables. The self-consistency of these pure particle theories is a non-trivial virtue, for theories according to which particles create fields that in turn act back on the particle are often beset by divergence difficulties.

The first fully Lorentz invariant particle theory was constructed by Henri Poincaré (1906). Poincaré sought an analogue of Newton's $1/r^2$ gravitational force law that would not only be formally Lorentz invariant but would also reflect the intuition that gravitational action is transmitted at the speed of light. The latter was taken to mean that for a particle $\#1$ moving under the gravitational influence of another particle $\#2$, the instantaneous acceleration of $\#1$ at the point (x_1^α, t_1), $\alpha = 1, 2, 3$ depends on the reciprocal of the square of the distance R_{21}

from #2 to #1 at the retarded time $t_2 = t_1 - R_{21}/c \, (= t_1 - R_{21}$ in our units). R_{21} must satisfy the functional relation

$$R_{21}^2(t) = \left(\sum_a (x_1^a(t) - x_2^a(t - R_{21}(t)))^2 \right)$$

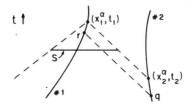

Fig. IV.6

Introducing four-vector notation, $U_\beta^i = dx_\beta^i/ds_\beta$ is the instantaneous four-velocity of particle β ($\beta = 1, 2$), where $x^4 = t$ and s_β is the proper time of β, related to the coordinate time by $ds_\beta = dt\sqrt{1 - v_\beta^2}$, $v_\beta = \sqrt{\sum_{a=1}^3 (dx_\beta^a/dt)^2}$. For the two particle system, Poincaré's equation of motion for particle #1 reads

$$\text{(IV.4)} \quad \frac{dU_1^i}{ds_1} = \frac{d^2 x_1^i}{ds_1} = -\frac{Gm_2}{\rho_2^3}\left(x_1^i - x_2^i - \frac{\rho_1 U_2^i}{\gamma} \right)$$

where x_1 and x_2 are evaluated respectively at t_1 and t_2 and where $\gamma \equiv \eta_{ij} U_1^i U_2^j$, and $\rho_\beta \equiv \eta_{ij}(x_1^i - x_2^i)U_\beta^j$. The equation of motion for #2 is similar. (Aside: Why does the strange looking term $\rho_1 U_2^i/\gamma$ appear on the right hand side of (IV.4)? It is a basic fact of Minkowski geometry that instantaneous four-velocity and acceleration and orthogonal, i.e., $\eta_{ij} U_\beta^i(dU_\beta^j/ds_\beta) = 0$. The reader should verify this constraint and show that (IV.4) meets it.) When the two particles are 'stationary' at the respective times t_1 and t_2 $((dx_\beta^a/dt)(t_\beta) = 0)$, then the spatial part of Poincaré's equation (IV.4) reduces to the Newtonian form

$$\text{(IV.5)} \quad \frac{d^2 x_1^a}{dt^2} = -Gm_2(x_1^a - x_2^a)/R_{21}^3$$

In 1911 de Sitter studied a more complicated version of Poincaré's equations in which the right hand sides contain a multiplicative factor of γ^n. By choosing n appropriately, the anomalous advance of the perihelion of Mercury can be accounted for. However, this choice of value for n does not allow for a consistent treatment of the gravitational red shift and it yields only one-half of the observed value for bending of light passing near the sun.[8]

Similar delay-differential equations occur in the relativistic electro-dynamics of charged particles. For two charged particles confined to one spatial dimension such that $x_1(t) < x_2(t)$, the equations are

$$(IV.6) \quad \frac{m_1 \dot{v}_1(t)}{[1 - v_1^2(t)]^{3/2}} = \frac{K[1 - v_2(t - R_{21}(t))]}{R_{21}^2(t)[1 + v_2(t - R_{21}(t))]}$$

$$\frac{m_2 \dot{v}_2(t)}{[1 - v_2^2(t)]^{3/2}} = \frac{K[1 + v_1(t - R_{12}(t))]}{R_{12}^2(t)[1 - v_1(t - R_{12}(t))]}$$

where $v(t) = \dot{x}(t)$, the dot denotes differentiation with respect to t, the constant K depends on the product of the charges, and the delays satisfy

$$(IV.7) \quad R_{21}(t) = |x_1(t) - x_2(t - R_{21}(t))|$$

$$R_{12}(t) = |x_2(t) - x_1(t - R_{12}(t))|$$

I will return to this case shortly since a good deal is known about the initial value problem. But before turning to the details, some general remarks about determinism in the present setting are in order.

First, particles which interact without the help of an intervening field medium violate the principle of action-by-contact and resist attempts to localize determinism. For the envisioned retarded action-at-a-distance mechanism, the state on a local slice, such as the S pictured in Fig. IV.6, cannot in general determine the state within $D(S)$. In the diagram, $r \in D^+(S)$, but the state on S cannot determine the state at r, for the retarded action from particle #2 reaches r without registering on S; since the postulated ontology contains only particle variables, there is no means by which to record the passage of the action on S. (This, by the way, tends to undermine the usefulness of the concept of domain of dependence, as we have defined it, for pure particle theories.)

More disturbing yet is the worry that no form of Laplacian deter-minism, local or global, will hold. To generate the worry, we need only consider a very simple time delay equation with constant delay; namely,

(IV.8) $\dot{x}(t) = -x(t - \pi/2)$.

The 'obvious' initial value problem is to set the value of x at $t = 0$, e.g. $x(0) = 0$. But this data does not determine a unique solution since, for instance, $A\sin(t)$ solves the initial value problem for any value of the constant A. Specifying $\dot{x}(0)$ fixes the value of the constant A for this solution, but there are still other solutions. Indeed, no matter how derivatives of x are given at $t = 0$, a unique solution is still not determined (Driver (1977)).

The next best thing to having the future determined by an infinitely thin slice of the history is to have it determined by a finite portion of the past history. Thus, let us suppose that for $t \in [-\pi/2, 0]$ $x(t)$ is given: $x(t) = \theta(t)$, where $\theta(t)$ is arbitrarily specified, without regard to whether it satisfies (IV.8), except for meeting the initial condition $x(0) = \theta(0)$. To find a continuous extension of θ into the future to a function satisfying (IV.8) we proceed in steps. In the first step we look at the interval $[0, +\pi/2]$. For this interval (IV.8) is reduced to the ordinary differential equation $\dot{x}(t) = -\theta(t - \pi/2)$, which can be solved explicitly. With this solution in hand we can iterate the procedure to extend the solution to the next interval $[+\pi/2, +3\pi/2]$, etc. (Driver (1977)).

Such examples give hope that the time-delay equations of special relativistic gravitation and electrodynamics are also deterministic in the near-Laplacian sense that a finite sandwich out of the past history determines the entire future. Of course, the physics problem is enormously more difficult than the artificial one from the preceding paragraph since we are now dealing with non-constant delays which depend upon the unknown solution. Not until the 1960's was this problem shown to have a positive outcome for electrodynamics, and then for the special case of two charged particles moving in one dimensional space. Driver (1963) showed that if the trajectories $x_1(t)$ and $x_2(t)$ are given for $t \in [\alpha, t_0]$, $-\infty < \alpha < t_0$, then there is a unique solution of (IV.6) for $t \in [t_0, \beta)$, $\beta > t_0$, which continuously extends the initial functions; and further, either $\beta = +\infty$ or else there is a collision as $t \to \beta^-$.

The hypotheses on which this result is proved are mostly unnoteworthy, technical conditions, save the one that asserts that at t_0 the functional relations (IV.7) for the delays are solvable. It might seem that this must always be attainable if α is made negative enough, but the must here is not the must of mathematical necessity as illustrated by the past hyperbolically accelerated particles of Fig. IV.7. Neither of the delay equations can be solved for $t \le t_0$. What this means physically, of course, is that no retarded influence from either particle can reach the other before t_0. We should thus give the system of equations (IV.6)—(IV.7) a more subtle reading. Consider candidate histories $x_1(t)$, $x_2(t)$, for $t \le t_0$, with $|v_1(t)|$ and $|v_2(t)|$ both less than c. If a solution of (IV.7) exists for $R_{21}(t)$ $(R_{12}(t))$, then the equation for $\dot{v}_1(t)$ $(\dot{v}_2(t))$ stands as stated in (VI.6); but if no solution for $R_{21}(t)$ $(R_{12}(t))$ exists, then the right hand side of the equation for $\dot{v}_1(t)$ $(\dot{v}_2(t))$ is to be set equal to zero. Under this interpretation the trajectories pictured in Fig. IV.7 do not pass muster since they do not satisfy $\dot{v}_1(t) = \dot{v}_2(t) = 0$ for all $t \le t_0$. If all other cases where either R_{21} or R_{12} fails to be defined could also be shown to fail to pass muster in this way, then we could restate the uniqueness result in the following form: assume that the reinterpreted (IV.6) is valid for all $t \le t_0$; then a finite portion of the past history uniquely determines the future. I would guess that this is so for the two particle case in general, but it is conceivable that with three or more particles moving under their mutual retarded actions we can obtain past hyperbolic acceleration, which in this setting is the unsettling analogue of the Newtonian particles that appear from spatial infinity (recall the discussion of Ch. III above).

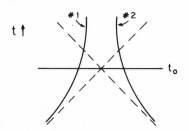

Fig. IV.7

More remarkably, Driver and coworkers have shown that despite what the cautionary examples like (IV.8) led us to expect, Newtonian initial data — instantaneous positions and velocities on a time slice — uniquely determine a solution for two like-charged particles (repulsive forces), again confined to one spatial dimension. The proof assumes that (IV.7) is solvable for all past times. It proceeds by first establishing that Newtonian data fix a unique past solution and then appeals to the previous result to conclude uniqueness for the future as well (Driver (1969); Hsing (1977)). There are some restrictions on the initial data (such as that the particles be sufficiently separated and are not approaching each other too quickly at the initial moment) but these are relatively mild. What one would like is for the assumption that (IV.7) is satisfied to be shown either to be dispensable or else to be justified along the lines suggested above. Otherwise it looks as if fiat is doing some of the work determinism should be doing.

Most remarkable of all, Driver (1978) has shown Laplacian style determinism to hold for some special cases of two-body motion when the retarded action is replaced by the time symmetric scheme of half-retarded plus half-advanced action. In particular, if two identically charged particles move symmetrically about the x-axis, they must come to rest at some moment, at which time they are at their minimum separation. Then if this separation is sufficiently large, Newtonian initial data determines a unique solution of the time symmetric version of (IV.6)—(IV.7).

Is Nature trying to tell us that Laplacian determinism will out in even the most unexpected cases, or are these results artifacts of the specialized problem sets — two particles, repulsive forces, one spatial dimension, sufficiently large initial separation, etc.? An indication towards the latter is given by numerical computations by Anderson and von Baeyer (1972) which suggest that in the symmetric half-retarded, half-advanced scheme, non-Newtonian degrees of freedom emerge if the minimum separation is small enough. But much more evidence is needed before a confident answer can be given. What is interesting for our purposes is that faith in Laplacian determinism is strong enough that there is no lack of suggestions for means for quashing the non-Newtonian degrees of freedom, should they exist (see Anderson and von Baeyer (1972)). For instance, a correspondence principle has been proposed which would rule out any solution which does not go over to a Newtonian solution as $c \to \infty$. This principle does its work, but in an unattractive way; for it starts from a lack of faith in the relativistic equations of motion and then proceeds to try to patch up these equations by appealing to another theory known to be false and superseded by relativity theory. But by now we know that there are no lengths to which true believers in Laplacian determinism are unwilling to go in order to preserve their belief. Still, it is somewhat disconcerting to realize that, just as in the Newtonian case, the faith was in place prior to the establishment of hard results for the Laplacian initial value problem.

8. INSTANTANEOUS ACTION-AT-A-DISTANCE

If instead of using delay terms we could write the relativistic equations of motion in Newtonian form, acceleration at an instant = function of positions and velocities at the same instant, then presumably Laplacian determinism would hold. Conversely, if instantaneous positions and velocities uniquely determine past and future trajectories, then we can write positions and velocities at a time as functions of the possible initial values. Differentiating the velocity function with respect to time gives acceleration as a function of initial values. Then solving the position and velocity relations for the initial values and inserting the result into the acceleration equation gives acceleration at t as a function of positions and velocities at t. The trouble, of course, with either of these directions is that it is not at all evident that Lorentz invariant equations of motion are attainable in the desired form.

The point is subtle. There is no difficulty in writing Lorentz invariant instantaneous action-at-a-distance equations; what is difficult is to assure both Lorentz invariance *and* a proper Laplacian initial value problem. Consider again the case of two particles. At the point p on the world line of particle #1 erect the instantaneous four-velocity U_1^i of #1 (see Fig. IV.8). Then draw the spacelike vector V^i which is Minkowski orthogonal to U_1^i ($\eta_{ij} U_1^i V^j = 0$) and which reaches to the world line of particle #2. At the point q where V^i touches #2 erect the four-velocity U_2^i of #2. The equation of motion for #1 will be assumed to have the form

$$(\text{IV.9}) \quad m_1 \frac{dU_1^i}{ds_1} = F_1^i(U_1^j, V^j, U_2^j)$$

where F_1^i is a four-vector concocted from the indicated arguments. For example, an analogue of Newton's $1/r^2$ law of gravitation is obtained by setting

$$(\text{IV.10}) \quad F_1^i = \frac{C m_1 m_2 V^i}{V^3} \qquad (V \equiv (\eta_{ij} V^i V^j)^{1/2})$$

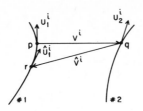

Fig. IV.8

From the point of view of the instantaneous rest frame of #1, only instantaneous quantities are involved. But now consider #2. Starting at the event q we draw the spacelike vector $\hat{V}^i$ which is orthogonal to U^i_2 and which reaches to the world line of #1. At the point r where $\hat{V}^i$ meets #1 we erect the four velocity $\hat{U}^i_1$ of #1. The equation of motion for #2 is of the form

$$(IV.11) \quad m_2 \frac{dU^i_2}{ds_2} = F^i_2(U^j_2, \hat{V}^j, \hat{U}^j_1)$$

By symmetry, we choose

$$(IV.12) \quad F^i_2 = \frac{Cm_1 m_2 \hat{V}^i}{\hat{V}^3}$$

When #1 and #2 are instantaneously at rest (U^i_1 and U^i_2 parallel), then $\hat{V}^i = -V^i$ and $\hat{U}^i_1 = U^i_1$, and the initial value problem appears to assume Newtonian form. But in general, data from more than one event on each of the particle trajectories seem to be needed to fix a unique solution.[9]

To get a surefire Laplacian initial value problem we can try another tack, now suppressing all but one spatial dimension for ease of presentation. In the (x, t) frame we try to write the ordinary accelera-

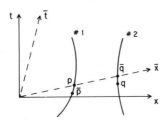

Fig. IV.9

tions $\dot{v}_1(p)$ and $\dot{v}_2(q)$ of the particles #1 and #2 at the simultaneous events p and q as functions of the instantaneous distance and velocity of the other particle:

$$(IV.13) \quad \dot{v}_1(p) = f_1(x_1(p) - x_2(q), v_1(p), v_2(q))$$
$$\dot{v}_2(q) = f_2(x_1(p) - x_2(q), v_1(p), v_2(q))$$

In the barred frame $(\bar{x}, \bar{t})$ p is simultaneous with $\bar{q}$ and q is simultaneous with $\bar{p}$ (see Fig. IV.9), so if the same lawlike relations hold in the barred frame as in the unbarred frame, we must also have:

$$(IV.14) \quad \dot{\bar{v}}_1(p) = f_1(\bar{x}_1(p) - \bar{x}_2(\bar{q}), \bar{v}_1(p), \bar{v}_2(\bar{q}))$$
$$\dot{\bar{v}}_2(q) = f_2(\bar{x}_1(\bar{p}) - \bar{x}_2(q), \bar{v}_1(\bar{p}), \bar{v}_2(q))$$

The Lorentz transformations give us the relations between $\dot{v}$ and $\dot{\bar{v}}$ taken at the same space-time point, and they also give the relation between p and q on one hand and $\bar{p}$ and $\bar{q}$ on the other. These relations in turn impose restrictions on the functions f_1 and f_2. These restrictions were obtained in differential form by Currie (1966) and independently by Hill (1967) and were later solved in implicit form by Hill (1970).[10] There are non-trivial solutions so that the Lorentz-invariant instantaneous action-at-a-distance formalism is at least mathematically consistent.

For the force functions which satisfy the Currie-Hill instantaneous action-at-a-distance scheme, we have a Newtonian type initial value problem and Laplacian determinism. But it is legitimate to ask whether we have determinism as anything more than a formal mathematical trick. The answer turns on several factors. It depends in the first place on what kinds of interactions can be accommodated in this formalism. Unfortunately, not enough is known here since only a few explicit force functions fitting the scheme have been constructed. The answer depends also on some tricky interpretation problems. For example, what happens if a bystander pokes particle #1 at event p? We seem to get the conflicting results that as viewed in the (x, t) frame the effect will be felt by #2 at event q, but as viewed in the $(\bar{x}, \bar{t})$ frame the effect on #2 won't be felt until the later event $\bar{q}$; and in either case we seem to have a signal traveling faster than light. As an initial response, it is fair to say that the equations of motion apply only to closed systems, and that they cannot be expected to yield consistent answers when the system is exposed to external perturbations. However, we should be able to add the source of the perturbation to the original system to obtain a combined and now closed system; but if the perturbing influence is to be described as a local interaction between the perturbing source and the particle in question, then exactly the same considerations as before suggest that a clash occurs with the non-local instantaneous action-at-a-distance description of the particle interactions. In this way Hill (1967a) was led to postulate that, for example, the measurement of the position of a particle cannot be described as a purely local interaction between the measuring instrument and the particle in question but involves a non-local interaction of the instrument with the other particles. Yet another suspicion arises when the scheme is extended to include more than two particles, for then it is found that the forces cannot be written as sums of two-body forces. Field theorists will see this holistic character of the forces as a vestige of the fields which have been suppressed in the pure particle description.

9. TACHYONS

We have assumed that faster-than-light propagation of mass-energy is a relativistic impossibility. The assumption seems safe on two sorts of grounds. First, it was part of the motivation for leaving Newtonian space-time for Minkowski space-time; thus, to challenge the assumption seems to imply a challenge to the relativistic conception of space and time. Second, once we are in Minkowski space-time, the assumption seems to be secured for both fields and particles. Lorentz invariant field equations, e.g., the scalar wave equation and Maxwell's electromagnetic field equations, propagate fields at the speed of light. For non-zero rest mass particles a simple calculation shows that an infinite amount of energy is needed to accelerate the particle from a speed $v < c$ to $v = c$; and for $v > c$ the various formulae involving factors of $\sqrt{1 - v^2/c^2}$ turn into gibberish.

The advocates of relativistic tachyons argue that there is a loophole through which these swift particles can fly. Tachyons don't have to be accelerated to a speed exceeding that of light since they always have and always will travel at superluminal speeds. The gibberish formulae which would result if, *per impossible*, slow moving tardyons could be accelerated to superluminal speeds are reworked and reinterpreted to tell a coherent story for tachyons. And finally, it is urged, no problem about relativistic simultaneity results since tachyons cannot be used to send 'messages' by which distant clocks could be non-relativistically synchronized.

I will not attempt to adjudicate the tangle of issues surrounding tachyons. All I want to do here is to emphasize that tachyons don't have to exist in the actual world to arouse new alarms for determinism. Determinism, remember, is a claim about all physically possible worlds, and Laplacian determinism seems to stumble over the mere physical possibility of tachyon flight. In the (x, t) frame of Fig. IV.10, the tachyon track ψ appears as an infinitely extended object which exists for but a fleeting instant. Since it has no history which registers on any t = constant slice before or after the one that contains it, both historical and futuristic Laplacian determinism seem to falter when initial data are given on an arbitrary t slice. In the barred frame $(\bar{x}, \bar{t})$ shown, ψ appears to be the track of an object which is moving swiftly from $\bar{x} = +\infty$ to $\bar{x} = -\infty$, and this object leaves its tracks on every $\bar{t} =$ constant slice. But determinism formulated with respect to this slicing is threatened by the tachyon track χ which in the barred frame is seen as a momentary object going nowhere and in the unbarred frame as an object moving swiftly from $x = -\infty$ to $x = +\infty$.

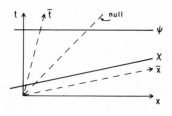

Fig. IV.10

Both to tolerate the physical possibility of tachyon flight and secure the possibility of Laplacian determinism for at least some slicings, we could seek to corral the tachyons so that they cannot roam the entire range but are confined to some finite speed $\hat{c} > c$. But to make that an invariantly meaningful restriction, we would need to add some additional structure to Minkowski space-time, e.g., perhaps describing tachyon flight by a second Lorentz signature metric $\hat{\eta}$ whose cones are 'wider' than the light cones of the Minkowski metric η describing tardyon behavior. Redefining 'spacelike,' 'domain of dependence,' 'Cauchy surface,' etc. to suit the tachyonic metric $\hat{\eta}$ restores the possibility of Laplacian determinism in tachyonic worlds. Exactly how physics is to be done in such a bi-metric setting remains to be seen, but clearly this setting carries us beyond the bounds of the orthodox special relativistic conceptions of space and time.[11]

To attain a peaceful coexistence between tachyons and determinism in orthodox Minkowski space-time other strategems could be tried. For example, the flight of the free tachyons ψ and χ in Fig. IV.10 could be curtailed by declaring against free tachyons and requiring that all tachyons are emitted by tardyonic matter. But what counts as a source and as an emission in one frame will appear as a sink and as an absorption in another frame. The requirement can be made symmetric and invariant by declaring against any loose tachyonic ends, i.e., that both ends of a tachyon world line terminate in tardyonic matter. But we will want to know how Nature manages so neatly to tie up loose ends, and in the absence of such an explanation we may begin to suspect that tachyons are merely a *façon de parler* device for talking about how tardyonic particles act upon one another by 'exchanging' swift particles.

Lacking a satisfactory way to establish peaceful coexistence, the determinist will see a strong argument against the possibility of tachyons; the tachyon enthusiast will see more evidence of the fragility of the doctrine of determinism.

10. CONCLUSION

From the perspective of determinism, special relativistic physics is emphatically not just 'more of the same.' For the first time we encounter examples where it is clear that any equivocation or shadow of turning in future events is ruled out by the combination of the present state of affairs and the laws of physics, unaided by fiat, stipulation or pious hope disguised as 'boundary conditions'. At least this is so for relativistic field theories. For particle theories the situation is less clear cut, for there remain open questions of mathematics and physical interpretation, both for the retarded and/or advanced and the instantaneous-action-at-a-distance approaches to relativistic particle mechanics. Tachyons threaten to undermine the toehold Laplacian determinism has thus far managed to establish, for these swift particles are the relativistic counterparts of the Newtonian influences which appear from or disappear to spatial infinity without portent or trace. The light cone structure of Minkowski space-time was initially assumed to shield against such surprises, but to the extent that the shield is porous, Laplacian determinism is just as much in jeopardy as it was in Newtonian worlds.

This chapter reconfirms a lesson learned in Ch. II; namely, we can't just read off the lesson for determinism from various branches of physics, for the implications we read will depend upon judgments about the adequacy of physical theories and those judgments will depend in turn on our views about determinism. This circularity need be no cause for alarm; indeed, it is what makes determinism a useful instrument for probing ontological and methodological problems. The probe is sometimes sharp, sometimes blunt, but nearly always guaranteed to locate something fundamental.

NOTES

[1] Einstein (1905). The paper is reprinted in English translation in Perrett and Jeffrey (1923).

[2] For some opinions on the matter, see Miller (1981) and Earman, Glymour and Rynasiewicz (1982).

[3] The reader interested in the details should consult Friedman (1983).

[4] For details, see Malament (1984).

[5] Some authors define $D^{\pm}(R)$ using timelike curves rather than causal curves. This difference makes only for a difference in the boundary points of $D^{\pm}(R)$.

[6] Popper (1982); recall the discussion from Ch. II above.

[7] For other intriguing examples where relativistic prediction is possible, see Geroch (1977).

[8] For a review of Poincaré type theories of gravitation, see Whitrow and Morduch (1965).

[9] 'Seem' because very little is known about existence and uniqueness of various forms of the initial value problem for equations such as (IV.9) and (IV.11).

[10] Hill's work was directly motivated by a desire to have a properly posed Laplacian initial value problem for relativistic mechanics; see Hill (1982).

[11] In this connection, see Nordtvedt (1974).

SUGGESTED READINGS FOR CHAPTER IV

For the historical background to STR, consult Miller's (1981) *Albert Einstein's Special Theory of Relativity*. There are literally dozens of textbooks on STR; three good ones, each with somewhat different virtues, are Pauli's (1958) *Theory of Relativity*, Synge's (1964) *Relativity: The Special Theory* and Møller's (1972) *The Theory of Relativity*. The initial value problem for the relativistic wave equation is treated in John (1982) *Partial Differential Equations*. Driver's (1977) *Ordinary and Delay Differential Equations* is a readable source of information about existence and uniqueness for delay equations in general, but the reader interested in the electrodynamics problem in particular will have to consult Driver's research papers. Kerner's (1972) *The Theory of Action-at-a-Distance in Relativistic Particle Dynamics* contains reprints of some of the basic papers on the relativistic instantaneous action-at-a-distance formalism.

CHAPTER V

DETERMINISM AND LAWS OF NATURE

> The problem is sometimes put in the form that we
> all distinguish between uniformities due to natural law
> and those which are merely accidentally true, 'histori-
> cal accidents on the cosmic scale'; if natural laws are
> just uniformities, how can this distinction be made? It
> seems to me foolish to deny (as some Humeans do)
> that such a distinction is made in common speech; but
> it also seems perfectly sensible to try to give the
> rationale for this distinction within the ambit of a
> constant conjunction view.
>
> (R. B. Braithwaite, *Scientific Explanation*)

We have made a start on understanding what properties laws of nature
must have if the world is to be deterministic, but nothing much has
been said about what laws of nature are, about what distinguishes laws
from non-laws. And, strictly speaking, we are in the embarrassing
position of having no examples to work with, for none of the examples
of so-called laws cited in previous chapters is truly a law since what is
asserted has proven to be false (and, by meta-induction, a similar fate
awaits every such example??). This realization need cause no undue
alarm if we are willing to apply to the history of science a Principle of
Respect, recommending that when we encounter a textbook example of
a 'law' we assume, unless there are specific contextual indications to the
contrary, that (1) the scientists of the period had good reason to believe
that what the 'law' asserts is true (or approximately true), and (2) the
scientists of the period were justified in believing that, if what the 'law'
asserts is true, then it does indeed express a law of nature. While I
agree with the spirit of this principle, I think that some caution is
required in applying it. In the young sciences it may be a struggle to
find any informative generalization that works tolerably well, and so the
standards of lawhood may be lax. We can avoid this problem by
looking only to the mature sciences for our examples. But in the mature
sciences the search for laws is constrained by the record of past
successes and failures; research scientists assume, consciously or not,
that the candidate laws must have a certain mathematical form, must

incorporate certain variables, must conform to certain symmetry and invariance principles, must reduce in special cases to the old 'laws', must mesh with 'laws' in allied fields, etc. Here opposing snares await us. One is the vulgar relativism of seeing the notion of law so inextricably tied to a scientific community, a research tradition, or whatever that only historical *reportage* is possible. The other is the arrogant abstractionism of supposing that an analysis of laws amounts to no more and no less than finding a core concept that cuts across every branch of science and every period in the history of science. I will be careful to avoid the snare of relativism, but I will knowingly step into a mild form of the abstractionism snare as it applies to modern physics. For my focus in this chapter is on the attempts of philosophers of science to provide an abstractive analysis of laws of physics. My main concern will not be so much with the rather thin character of these attempts as with the discordance which has recently grown to the extent that it cannot be ignored. While unanimity is an unattainable and even undesirable goal in philosophy, something is amiss when we cannot agree even approximately on how to understand a notion that is fundamental to the study not only of determinism but to the methodology and content of the sciences in general.

When in doubt it is a good practice to return to the source. In this case the source is David Hume.

1. HUME'S DEFINITIONS OF 'CAUSE'

Hume defined 'cause' three times over. (Recall: The constant conjunction definition says that a cause is "an object precedent and contiguous to another, and where all the objects resembling the former are plac'd in a like relation of priority and contiguity to those objects, that resemble the latter." The felt determination definition takes a cause to be "an object precedent and contiguous to another, and so united with it in the imagination, that the idea of the one determines the mind to form the idea of the other, and the impression of the one to form a more lively idea of the other."[1] And finally, in the *Enquiry*, but not in the *Treatise*, Hume defines a cause as "an object followed by another ... where, if the first object had not been, the second never had existed."[2])

The two principal definitions (constant conjunction, felt determination) provide the anchors for the two main strands of the modern empiricist accounts of laws of nature[3] while the third (the counter-factual definition) may be seen as the inspiration of the non-Humean necessitarian analyses. Corresponding to the felt determination defini-tion is the account of laws that emphasizes human attitudes, beliefs, and actions. Latter day weavers of this strand include Nelson Goodman, A. J. Ayer, and Nicholas Rescher. In *Fact, Fiction and Forecast* Goodman writes: "I want only to emphasize the Humean idea that rather than a sentence being used for prediction because it is a law, it is called a law because it is used for prediction . . ." (1955, p. 26). In "What Is a Law of Nature?" Ayer explains that the difference between 'generalizations of fact' and 'generalizations of law' "lies not so much on the side of facts which make them true or false, as in the attitude of those who put them forward" (1956, p. 162). And in a similar vein, Rescher maintains that lawfulness is "mind-dependent"; it is not something which is discovered but which is supplied: "Lawfulness is not found in or extracted from the evidence, but it is superadded to it. *Lawfulness is a matter of imputation*" (1970, p. 107). By contrast, the constant conjunction definition promotes the view that laws are to be analyzed in terms of the *de re* characteristics of regularities, independently of the attitudes and actions of actual or potential knowers.

Hume himself gives passing acknowledgement to the fact that the two strands can diverge.[4] And where they diverge, I follow the constant conjunction strand and declare my starting assumptions that whatever our beliefs, we could be mistaken because there is something to be mistaken about — the distinction between uniformities due to natural laws and those which are merely cosmic accidents is to be drawn in terms of features of the uniformities and not in terms of our attitudes towards them.[5] At the same time I readily concede that this strand cannot be successfully woven into an account of laws by completely ignoring the other strand, for while ontology need not follow epis-temology, our account of laws must explain how it is possible to form rational beliefs about what the laws of our world are. The hope is that this epistemological constraint can be met without becoming so entangled in the felt determination strand that we become captives of the Goodman-Ayer-Rescher web.

Against this hope I sense a rising sentiment among philosophers of science that the problem of giving a regularity analysis of laws bears an ominous resemblance to the problem of providing a criterion of 'cogni-tive significance' to separate empirically meaningful assertions from metaphysical nonsense. It was initially an article of faith among the positivists and logical empiricists that such a criterion must exist and that providing it in a suitable form was only a matter of finding the appropriate technical formulation. But as attempt after attempt fell into the philosophical waste bin this faith has given way to an indifferent agnosticism or, worse, an insipid lip service. If a similar ignominious fate awaits the regularity account of laws, then it would seem best to redirect our efforts elsewhere.

A growing band of philosophers is already at work in the elsewhere, constructing a non-empiricist conception of laws. But before turning to their views, let us review the sources of dissatisfaction with the standard regularity account and explore the prospects of improving it within an empiricist framework.

2. THE NAIVE REGULARITY ACCOUNT

The crudest form of the regularity account puts laws of nature and Humean regularities into one-one correspondence. In the linguistic mode favored by the logical positivists, this account might be rendered thus:

(H)   Laws are what are expressed by true lawlike sentences.

What makes the naive regularity account naive is the assumption that 'lawlike' can be captured by syntactical and semantical conditions on individual sentences. E.g., S is lawlike just in case S is general in form (say, a universal condition $(x)(Fx \supset Gx)$ so dear to philosophers determined to make use of their required symbolic logic course) and the predicates are suitably kosher ('F' and 'G' are non-positional, purely qualitative, non-Goodmanized, etc.) This is, to be sure, sloppy and vague, but the impression given by the older references was that all the mysteries of laws would disappear once the appropriate technical apparatus was applied to make notions like 'generality' and 'non-positional predicate' really precise.[6]

We do not need to await the outcome of the technical maneuvers. W. A. Suchting, David Armstrong, and other down-under philosophers have done such a thorough demolition job on the naive regularity account that we can be confident that no way of fiddling with the details of (H) will produce a defensible version. I will just remind you of some of the considerations and refer you to Armstrong (1983) for further details.

There is first the difficulty of uninstantiated lawlike generalizations. To exclude all such generalizations from law status is too severe; witness Newton's First Law ("If the net impressed force acting on a massive body is zero, then the body moves inertially") whose antecedent is very unlikely to be instanced in a universe well populated by massive particles obeying Newton's Law of Universal Gravitation.[7] Contrariwise, to welcome in all uninstanced lawlike generalizations has even more unwelcome consequences, for then the vacuity of the antecedent condition would mean that $(x)(Fx \supset Gx)$, $(x)(Fx \supset G'x)$, $(x)(Fx \supset G''x)$, etc., where Gx, $G'x$, $G''x$ etc., may be pairwise incompatible, are all laws. Such contrary 'laws' are intuitively repugnant, and they pose difficulties for the widely accepted view that laws license subjunctive conditionals. If o (which as a matter of fact is non-F) were F, would it be G, or G', or G'', etc.? A uniform treatment of uninstanced generalizations is unacceptable. But what basis does the naive regularity theorist have for treating such generalizations differentially?

The problem of uninstanced generalizations pales beside the problem of instanced lawlike generalizations which, by the judgments of philosophical intuition and the history of science, do not correspond to laws. Reichenbach's old example still suffices: "All bodies of pure gold have a mass of less than 10,000 kg." This statement is general in form; its predicates are surely kosher; and it is widely instanced. But even if we were assured that it is true, we would not regard it as expressing a law. Nor would it help to be given the further assurance that the known instances are not exhaustive or that there are an infinite number of instances (so that the generalization is not equivalent to a finite conjunction of singular statements). Such assurances would do nothing to convince us that Reichenbach's generalization is a generalization of law rather than of fact.

Can the separation of generalizations of law from generalizations of fact be effected by *de re* features of regularities, or as empiricists are we forced to grasp the safety cord of Hume's felt determination definition? My strategy for answering this question will be, first, to state general constraints on an empiricist account of laws and, second, to explore the prospects and problems of constructing a more appealing regularity account within the confines of these constraints.

3. THE EMPIRICIST CONSTRAINTS

I will state the constraints in a form that may be distasteful to some empiricists. But to mix a metaphor, while I can genuflect before Hume's altar with the best of them, I am no knee-jerk empiricist. I see no reason to deny ourselves whatever analytical tools may help to shape the issues into a manageable form. Without further apology, I state the basic or 0-th empiricist constraint as

(E0)   Laws are contingent, i.e., they are not true in all possible worlds.

Next, I propose two forms for further constraints:

(F1)   For any possible worlds W_1, W_2, if W_1 and W_2 agree on ___, then W_1 and W_2 agree on laws.

(F2)   For any possible worlds W_1, W_2, if W_1 and W_2 agree on laws, then W_1 and W_2 agree on ___.

The blanks are to be filled in by non-question-begging empirical features. 'All Humean regularities' is such a feature, but if used as the filling in both blanks it seems that the conjunction of the resulting constraints forces us back to the naive regularity account.

The filling I prefer for the blank in (F1) produces the following constraint:

(E1)   For any W_1, W_2, if W_1 and W_2 agree on all occurrent facts, then W_1 and W_2 agree on laws.

I will refer to (E1) as the empiricist loyalty test on laws, for I believe it captures the central empiricist intuition that laws are parasitic on occurrent facts. Ask me what an occurrent fact is and I will pass your query on to empiricists. But in lieu of a reply, I will volunteer that the paradigm form of a singular occurrent fact is: the fact expressed by the sentence $P(o, t)$, where 'P' is again a suitably kosher predicate, 'o' denotes a physical object or spatial location, and 't' denotes a time. (This is a qualitative version of one of the classical world assumptions used earlier (see Ch. III).) There may also be general occurrent facts (I think there are), but these presumably are also parasitic on the singular occurrent facts. Conservative empiricists may want to restrict the antecedent of (E1) so as to range only over observable facts while more liberal empiricists may be happy with unobservable facts such as the fact that quark q is charming and flavorful at t. In this way we arrive at many different versions of the loyalty test, one for each persuasion of empiricist.

The well-known motivations for (E1) fall into two related categories. There are ontological argument and sloganeering ("The world is a world of occurrent facts"), the two often being hard to distinguish. Then there are epistemological arguments and threatenings, the most widely used being the threat of unknowability, based on two premises: we can in principle know directly or noninferentially only (some subset of) occurrent facts; what is underdetermined by everything we can in principle know non-inferentially is unknowable in principle. I will return to this argument in Sec. 12 below. The argument connects back to the ontological if we add the further premise that what isn't knowable in principle *isn't* in principle.[8]

Finding a filling for the blank in (F2) which produces a defensible but not toothless constraint is more difficult. Consider:

(E2)   For any W_1, W_2, if W_1 and W_2 agree on laws, then W_1 and W_2 agree on regularities entailed by the laws.

This lacks bite in the case of non-probabilistic laws, but it is of some help in separating some of the views on the nature of physical probabilities. Hardcore frequency theorists would hold if W_1 and W_2 agree on lawful probabilities and if they both contain infinite repetitions of the relevant chance experiment, then they must agree on limiting relative frequencies; but the hardcore propensity theorist will counter that while agreement of relative frequencies is likely, it is not mandatory. However, a more important difference between frequency and propensity theorists concerns (E1) and the grounding of physical probabilities on occurrent facts (see Sec. 9 below and Ch. VIII). Little use will be made of (E2) in what follows.

Two things remain uncaptured by (E0)–(E2). Neither can be stated in the form of a tidy constraint, but nonetheless each is an important part of the empiricist conception of laws. The first is the intuition that appropriate qualitative and quantitative differences in particular occurrent fact and general regularity make for differences in laws (E3). The second intuition is that there is a democracy of facts and regularities in that each has a vote in electing the laws (E4). The worry about (E4), of course, is whether democracy can prevail without degenerating into the mob rule of the naive regularity view. And the problem with (E3) is that it seems impossible to specify ahead of time in a content and context free manner what counts as an appropriate difference. That (E3) and (E4) are painfully vague does not mean that they are useless; on the contrary, a good check on any proposed implementation of (E0) and (E1) is how well it makes sense of (E3) and (E4).

In the next section I will review what I take to be the most promising approach to laws which fulfills the above constraints and which maintains firm contact with Hume's constant conjunction idea. I will capitalize the *e* in 'empiricism' to indicate my brand of empiricism. There are other and perhaps better brands, but this one recommends itself as a useful foil.

4. MILL, RAMSEY, AND LEWIS

John Stuart Mill, as thoroughgoing an Empiricist as they come, was no naive regularity theorist. Humean uniformities are often called laws in common parlance; but scientific parlance is quite another thing:

Scientifically speaking, that title [Laws of Nature] is employed in a more restricted sense to designate the uniformities when reduced to their most simple expression. (1904, p. 229)

This 'restricted sense' is explained more fully a little further on:

According to one mode of expression, the question, What are laws of nature? may be stated thus: What are the fewest and simplest assumptions, which being granted, the whole existing order of nature would result? Another mode of starting the question would be thus: What are the fewest general propositions from which all the uniformities which exist in the universe might be deductively inferred? (1904, p. 230)

When allowance is made for the fact that Mill assumed determinism, his conception of laws seems to correspond exactly to Frank Ramsey's, or rather to David Lewis' de-epistemologized version of Ramsey. Ramsey's dictum was that laws are "consequences of those propositions which we should take as axioms if we knew everything and organized it as simply as possible in a deductive system" (1978, p. 138). Lewis suggests we expunge the reference of knowledge in favor of conditions on deductive systems, known or unknown: ". . . a contingent generalization is a law of nature if and only if it appears as a theorem (or axiom)

in each of the true deductive systems that achieves a best combination
of simplicity and strength" (1973a, p. 73). Deductive systems are

deductively closed, axiomatizable sets of true sentences. Of these true deductive
systems, some can be axiomatized more *simply* than others. Also some of them have
more *strength*, or *information content*, than others. The virtues of simplicity and
strength tend to conflict . . . What we value in a deductive system is a properly balanced
combination of simplicity and strength — as much of both as truth and our way of
balancing will permit. (1973a, p. 73)

Many other forms of the idea that lawhood attaches to individual
regularities only via their membership in a coherent system of regularities could be cited,[9] but for the moment let us stick with the
Mill-Ramsey-Lewis version and enumerate its virtues.

I take it as evident that the M-R-L account does satisfy the basic
Empiricist constraints (E0) and (E1), does provide for the democracy
of facts and regularities (E4) without surrendering to the mob rule of
the naive regularity account, and does provide a framework for understanding what sorts of differences in particular fact and general
regularity make for differences in laws (E3). It also has the virtue of
explaining why laws have or tend to have various 'lawlike' characteristics, such as universality (more on this in Sec. 6 below). It allows in
some vacuous generalizations without opening the floodgates to all.
And it connects in a direct and natural way to the actual practice of
scientific theorizing or at least to the most widely held reconstruction of
the practice in the form of the hypothetico-deductive method. In fact, in
much of the current literature on the structure and function of scientific
theories, 'theory' and 'deductive system' can be freely interchanged.

5. DEDUCTIVE SYSTEMATIZATION: A CLOSER LOOK

It is no criticism of M-R-L to note that simplicity and allied notions
such as coherence and systematization are vague and slippery, for so is
the notion of laws of nature. The question is whether the vaguenesses
and slippages match. That old nemesis, Reichenbach's gold lump
generalization, gives pause. If this generalization is to be counted out as
a law by the lights of M-R-L it is because it is not an axiom or theorem
in the best (or each of the best) overall deductive systems for our world.
Consider then what would happen if we were to add it as an additional

axiom. There would, by hypothesis, be a gain in strength. And, presumably, there would also be a loss in simplicity. The loss must, *pace* M-R-L, outweigh the gain. I will not say otherwise. But I do say that it is not compellingly obvious that the scales tip in this way while it is compelling that Reichenbach's generalization is not to be counted as a law.

The trouble here may not lie with the squishy notion of simplicity but with the seemingly more solid notion of strength. Lewis suggests strength be measured by information content, and that is as good a measure as any if we are interested in strength *per se*. But actual scientific practice speaks in favor not of strength *per se* but strength in intended applications; for dynamical laws this means strength as measured by the amount of occurrent fact and regularity that is systematized or explained relative to appropriate initial and/or boundary conditions. The advantage offered by deterministic generalizations here is obvious: while they can be strengthened *per se*, they are, in their intended applications, as strong as strong can be; for given the state of the system at any instant, they entail everything true of the system, past, present, and future, and any other generalization is either incompatible or adds nothing to applied strength. This helps to explain why we feel confident that in having discovered a simple set of true deterministic relationships we have discovered laws. This is not to say that determinism is either necessary or sufficient for a good trade-off between simplicity and applied strength. If a deterministic set of generalizations can be constructed only at the price of very high complexity, then the scales may tip against determinism; but typically the complexity must be great indeed before the tip becomes pronounced. And when no set of true deterministic generalizations is available, many different compromises between simplicity and strength may recommend themselves. This helps to explain why, independently of ontological considerations, determinism has been prized as a methodological guide to scientific theorizing.

What started as an objection to the M-R-L account has turned into a plus. Another plus comes from reflection on the notion of chaos. It is tempting to define chaos as the absence of any pattern or regularity, but the discussion in Ch. VIII will cast doubt on the coherence of this idea. However, chaos as the non-existence of laws is explicable on the M-R-L account. This form of chaos need not require that all regularity is absent but only that the existent regularities are sufficiently weak and messy that there are no good compromises between strength and simplicity.

In closing, I have to confess to a real worry about the M-R-L account, or rather to the linguistic version I have been reviewing. Given a choice of language — primitive predicates and logical apparatus — we may be able to identify a best overall deductive system. But different choices of language may promote different candidates for the role of best system. These candidates may be incommensurable, not admitting meaningful comparisons of simplicity and strength. Or else they may be commensurable and equally good in their different ways, forcing us to say either that there are no laws since there are no non-trivial axioms or theorems common to all the best systems, or that the laws are relative to a choice of language. These worries can be diminished by refusing to give in to the logical positivists' fear of the ontological and their flight to the linguistic. Recall that my canonical formulation of determinism assumes that the possible worlds can be characterized in terms of space-time magnitudes. Worlds are thus isomorphic to sets of basic propositions, each asserting that the value of such-and-such a magnitude takes a value of so-and-so at thus-and-such a spatio-temporal location. The laws of the actual world are then the propositions that appear in each of the deductively closed systems of general propositions that achieve a best systematization of the basic propositions true of the actual world. So while different systems may employ different concepts, there will of necessity be a strong common core.

6. LAWS AS UNIVERSAL AND ETERNAL TRUTHS

Different philosophers of science draw up different lists of properties they think laws should have, but there is wide-spread agreement that laws should express universal and eternal truths; that is, they should apply unrestrictedly to all space and all time, and they should not change with time.

The notion that laws 'change with time' is ambiguous as between having a single law that is not time translation invariant vs. having different laws in different epochs. I can offer no general criterion to decide which case is which. What I can offer for the former category is an analysis of the meaning of time translation invariance, its status as a requirement of lawhood, and its connection with other symmetry principles and with determinism. This I will do in Ch. VII. For cases which fall into the latter category we obviously have a violation of universality since the putative laws apply only to limited stretches of time.

No purely syntactical criterion can capture the intended sense of universality. If we are clever enough we can always rephrase any assertion so that it has the form $(x)(t)[\ldots]$, which appears to assert something about *all* space and *all* time. The appearance can be belied by the intended meaning of the predicates and relations which fill the ellipsis. To cite the standard example, $(x)(t)[P(x, t) \supset S(x, t)]$ appears to be general in form and universal in scope, but if '$P(x, t)$' means that x is in Nelson Goodman's left-hand trouser pocket for t between March 30 and 31 in 1948, the relevant sense of universality is absent. It will not do, however, to exclude all predicates and relations whose intended interpretations refer to particular spatio-temporal regions; we can, for instance, formulate a relationship between the rate of expansion of the universe and its mass content which refers explicitly or implicitly to the initial 'big bang' but which applies to all space-time.

For a space-time theory in my sense we can easily express the desired sense of universality in model-theoretic terms. Recall that the intended models have the form $\mathcal{M} = \langle M, 0_1, 0_2, \ldots \rangle$ where M is the space-time manifold and the 0_i are geometric object fields on M. For any subregion $R \subset M$, the restriction $\mathcal{M}|_R$ of $\mathcal{M}$ to R is $\langle R, 0_1|_R, 0_2|_R, \ldots \rangle$. The putative law L of T lacks universality just in case it does not apply to some non-empty R; intuitively, as far as L is concerned, 'anything goes' in R. We can take this to mean that for any logically possible $\mathcal{M}, \mathcal{M}|_R$ satisfies L.

The M-R-L account of laws explains why universality is prized as a feature of laws; namely, it promotes both strength and simplicity. Still, is it conceivable that the occurrent regularities of the world could be so structured that there is an obviously best overall compromise between strength and simplicity involving 'laws' that are not universal? I will leave it to the reader to construct and evaluate examples for herself. A positive result will be counted against the M-R-L account by those who promote the relations-among-universals view of laws since for them laws are necessarily eternal and universal.

7. DEFEASIBILITY AND DEGREES OF LAWFULNESS

Mill has been read as defining laws as indefeasible or unconditional uniformities. This is, I think, a backwards reading. 'Unconditional' is analyzed as "invariable under all changes of circumstance," but the range of circumstances that may serve as defeasors is defined to be precisely those allowed by the "ultimate laws of nature (whatever they may be) as distinguished from the derivative laws and from the collocations" (Mill, 1904, p. 244). And these ultimate laws are defined as the axioms of the M-R-L system. Thus, it is only when we have in hand some candidate for the M-R-L axiom system that a defeasibility analysis can begin.[10]

Ramsey had a similar idea in distinguishing four categories of universal generalizations. At the top are the ultimate laws; then come, in descending order, derivative laws, then those called laws "in a loose sense," and finally the universals of fact. Derivative laws are simply the universal generalizations that are theorems of the best deductive system. Laws-in-a-loose-sense are those general propositions deducible using "facts of existence assumed to be known by everybody" (1978, p. 130). Universals of fact are the accidental or non-lawful generalizations. Ramsey was quick to note that the last two categories cannot be sharply separated. The separation rests on the amount of fact allowed in the deduction; if, for example, determinism is true, all universals of fact can be deduced from the ultimate laws together with enough facts of existence. (This is why Mill, a determinist, defined the ultimate laws to be the fewest general propositions from which *all* the uniformities which exist in the universe are deducible.) Nevertheless, I agree with Ramsey that the distinction is a useful one, and I propose to redraw it in a somewhat more elaborate form and relate it to Mill's defeasibility notion.

Ideal and complete defeasibility of a universal generalization of fact would show how its truth or falsity turns on contingencies by providing (i) a two-fold partition of the initial/boundary conditions compatible with the M-R-L axioms into those which together with the axioms guarantee the failure of the generalization (the defeasors) vs. those which together with the axioms guarantee the truth of the generalization (the enablers), and (ii) a demonstration of how generic or exceptional the enablers and defeasors are in the models of the axioms. The suggestion then is that degrees of indefeasibility or lawfulness can be assigned depending on the results of (ii). Those generalizations whose enablers are exceptional ('measure zero') and whose defeasors are generic are rightly called merely accidental, while those whose enablers are generic and whose defeasors are exceptional can approach lawhood.

Such a classification scheme demands much of our laws — (i) presupposes that the ultimate laws are appropriately complete, and (ii) assumes that a suitable measure can be defined on the initial/boundary conditions — and there is no *a priori* assurance that nature will answer these demands. But as an example where such a defeasibility analysis has been carried out, I would cite recent work on the singularities of gravitational collapse as described by Einstein's general theory of relativity. The singularities first discovered in solutions to Einstein's field equations were thought to depend on the idealized features of the special models under study (viz., perfect spherical symmetry). However, several decades of work, culminating in the theorems of Penrose and Hawking showed that the situation is just the reverse; singularities of collapse develop under generic conditions (see Ch. X).

8. CHALLENGES TO THE REGULARITY ACCOUNT OF LAWS

The details of my Empiricist constraints will be filled in in different ways by Empiricists of different stripes. And once the details are supplied, it remains to settle on the best means of implementing the key constraint (E1) by specifying how the occurrent facts determine the laws; perhaps, as argued, a slightly modified version of M-R-L is the best bet, perhaps not. All of this is subject to continuing discussion and debate. But one wonders whether the basic thrust of the Empiricist program as I outlined it is seriously discussible, or whether any attempt at discussion quickly degenerates into an exchange of Empiricist and anti-Empiricist epithets. In what follows I will try to describe challenges to the Empiricist account in such a way that something can be learned from the resulting debates, though eventual termination in irreconcilable intuitions is to be expected.

I will postpone until Ch. XI a discussion of the challenge the quantum theory poses to the occurrent ontology presupposed in the classical world view and in the Empiricist constraint (E1). However, non-occurrent dispositions, potentialities, and propensities are all encountered outside of the strange realm of quantum physics. Their implications for a regularity account of laws will be taken up next. I will then turn to the recent attack which has been launched against the regularity account by Armstrong, Dretske, Tooley and others who see laws of nature as relations among universals.

9. DISPOSITIONS: THE GARDEN VARIETY TYPE

I want first to clear away two sources of misunderstanding about dispositions, due ironically, to two of the most important contributors to the subject, Rudolf Carnap and Sir Karl Popper.

On Carnap's analysis of scientific concepts, disposition terms occupy an "intermediate position between observation terms . . . and theoretical terms" (1964, p. 63). Given Carnap's epistemologically oriented concern with testability and meaning, it is easy to appreciate the motivation for his classification, but for our purposes the ontological dimension is more important. Along this dimension, the contrast to dispositional is neither observational nor theoretical but occurrent. And the dispositional-occurrent distinction does not lie parallel to the theoretical-observational distinction but oblique to it — theoretical terms may denote either occurrent or dispositional properties (see Mellor (1971) Ch. 4).

My other complaint is aimed at Sir Karl's contention that

all physical . . . properties are dispositional. That a surface is coloured red means that it has a disposition to reflect light of a certain wave length. That a beam of light has a certain wave length means that it is disposed to behave in a certain manner if surfaces of various colours, or prisms, or spectrographs, or slotted screens, etc., are put in its way. (1962, p. 70).

Taken literally, Sir Karl's remarks threaten to erase occurrent ontology, leaving a squirming, twisting mass of dispositions. On further reflection, however, it is clear that the sense in which all physical properties are dispositional is a harmless guilt-by-association sense. As far as we know, each physical property is joined to others by lawful regularities. (And if this were not so, how could we know it was not so?) This is especially true of properties denoted by theoretical terms because these terms are usually introduced precisely for the purpose of formulating laws. Thus, that a beam of light has a certain wave length does 'mean that' the beam is disposed to behave in a certain manner whenever a slotted screen is put in its way; that is, the relevant laws of optics license the subjunctive conditionals about such behavior. But having a given wave length is unlike having a pure dispositional property (say, solubility) in two related respects: the former is an occurrent property and it has, in Carnap's phrase, an 'open texture' in that its meaning is not exhausted by any one or even a collection of such subjunctive conditionals.

Garden variety dispositions, like solubility, hardly require Empiricism to flex its muscles. We are confident that the secrets of dispositions to dissolve are to be found jointly in (a) occurrent facts about the micro-structure of salts and crystals and (b) laws couched purely in terms of occurrent properties. Thus, to the extent that we are convinced that the relevant laws pass the Empiricist loyalty test (E1), we can likewise be confident that dispositions to dissolve do not hold non-Humean powers:

(D)   For any W_1, W_2, if W_1 and W_2 agree on all occurrent facts, then W_1 and W_2 agree on dispositional facts regarding solubility (and other garden variety pure dispositions).

Nothing in the more impassioned defenses of dispositions — such as Mellor's (1974) — moves me to abandon my Victorian prejudice. Garden variety dispositions, like unmarried mothers, cannot manage (it) on their own. And the success of science in showing how it is managed on an occurrent basis makes claims to the contrary seem like so much mystery mongering.

If it is mysteries you want they are ready and waiting for us once we move from garden variety dispositions to physical probabilities construed as probabilistic dispositions or propensities. I reject out of hand the view of the finite frequentists who identify physical probabilities with ratios in finite classes or sequences.[11] This view fits the most stringent form of Empiricism imaginable, but it makes analytically false any assertion which sets probability equal to an irrational number and also any assertion that makes the probability of the outcome of an experiment p, for p strictly between 0 and 1, when as a matter of fact the experiment is performed only once. Limiting relative frequencies in infinite sequences of outcomes escape these difficulties, but, to complete the march towards probabilities as dispositions, we need only add that an actually infinite repetition of the relevant experiment is rarely, if ever, to be found in nature.

Both the hypothetical frequency theorist and the propensity theorist agree that physical probabilities are, in some sense, dispositions or propensities.[12] The difference, roughly speaking, is that the frequency theorist is guided by a determination to remain Empiricist and tends to see probabilistic dispositions as reducible in much the same way we supposed the disposition of solubility to be. The most radical of the propensity theorists — those who assign probabilities to single cases — resist reductionism and sail close to the conclusion that propensity probabilities are non-Humean powers.

These issues will be developed in Chs. VIII and XI. In the remainder of this chapter I will restrict attention to non-probabilistic laws.

10. TOOLEY'S CASE

David Armstrong, Fred Dretske, Michael Tooley, and Chris Swoyer have all proposed that laws of nature are relations among universals. Their views differ in interesting and subtle ways, but for the moment I will lump Armstrong, Dretske, and Tooley together because they accept the minimal Empiricist constraint (E0) while Swoyer does not. The triumvirate is also unanimous in rejecting any form of the regularity account. As Armstrong puts it, "I am saying that we can keep the Humean uniformities fixed, and vary the laws indefinitely" (1983, p. 71, fn. 3). This is a disavowal not of (E1) but of the stronger

(E1′)   For any W_1 and W_2, if W_1 and W_2 agree on all Humean
          regularities, then W_1 and W_2 agree on laws.

But unless I misread Armstrong he intends to reject (E1) as well, as I think he must if he wants to overthrow every variant of M-R-L.

In trying to understand the intuition behind the rejection of (E1) it is useful to review a hypothetical case constructed by Tooley.

Imagine a world containing ten different types of fundamental particles. Suppose further that the behavior of particles in interaction depends upon the types of the interacting particles. Considering interactions involving two particles, there are 55 possibilities with respect to the types of the two particles. Suppose that 54 of these possible interactions have been carefully studied, with the result that 54 laws have been discovered, one for each case, which are not interrelated in any way. Suppose finally that the world is sufficiently deterministic that, given the way particles of types X and Y are currently distributed, it is impossible for them ever to interact ... In such a situation it would seem very reasonable to believe that there is some *underived* law dealing with the interaction of the particles of types X and Y ... (1977, p. 669)

Tooley argues further that the best M-R-L system for his hypothetical world will not contain any axioms or theorems describing how the unsociable particle species X and Y would behave if they were to interact. Something has to give, and what gives, according to Tooley, is the M-R-L account. It is then but a short step to the conclusions that no form of the regularity account will work and that if we want truth makers in the world for the underived laws about $X-Y$ interactions, we had best begin looking for relations among universals.

Leaving until later a discussion of the merits of the universals view of laws, I want to respond to the attack on M-R-L. I begin by asking Tooley how he can be so sure that the seemingly unsociable particle species are not acting upon one another at-a-distance and that the regularities of this interaction do not show up in the best overall deductive system. Or how can he be so sure that there is no unified particle theory which explains all ten species in terms of a more fundamental particle (the quack, say, which comes in ten honks) and which shows up as part of the best deductive system? The story can be told in increasing detail so as to rule out these and other such possibilities. But, the Empiricist would contend, the more such detail, the more implausible it becomes that there is any truth to the matter of laws about $X-Y$ interactions.

Part of the intuitive appeal of Tooley's example comes from the meta-induction he invites us to make on the basis of the 45 laws derived from observations of interactions of pairs of particles from different species. I agree that such meta-inductions can override initial first-order inductions where we build a M-R-L system on the basis of a limited range of observed regularities. But in the limit where the basis expands to include all occurrent facts and regularities, the meta-induction must give way to the first-order induction. For example, a meta-induction on derived laws may speak strongly in favor of some conservation principle, and if this principle clashes with the results of initial attempts to incorporate a newly discovered interaction into the best deductive system, then the meta-induction may prevail, sending the deductive system back to the drawing boards. But 'back to the drawing boards' means collecting more information in the form of occurrent facts and building a new deductive system on the broader basis. If no matter how far we expand the basis, the axioms still conflict with the conservation principle, then it is the meta-principle which must go.

The sophisticated Empiricist has no blanket prohibition against uninstantiated laws; such laws may well emerge as consequences of the axioms of the best deductive system. The unalterable Empiricist constraint is that such laws must perform in the service of the actual; they must arise in the attempt to account for actual occurrent facts and regularities. To hold otherwise is to break the grip the actual holds on the possible. It is easy to imagine richer and more sociable worlds where the species X and Y do interact and that from the regularities of interaction we derive the laws. The trouble is that it is all too easy, for we can imagine many such worlds, each with different M-R-L laws of $X-Y$ interaction. Armstrong's way of putting the point seems to me exactly right (see (1983), Ch. 8, Sec. 4). We do have the intuition that if conditions were such as to permit $X-Y$ intercourse, there would be some sort of $X-Y$ interaction laws. That intuition is, of course, perfectly consistent with M-R-L if in each of the possible worlds where intercourse is consummated, the regularities of interaction find their way into the best deductive system. But there is, in this world, no truth to the matter as to what form the laws of $X-Y$ interaction would take; for by Tooley's construction there is nothing in actual occurrent fact or regularity to allow us to say which of the possible worlds with $X-Y$ intercourse is 'nearest' or 'most similar' to this world.

11. NOMIC NECESSITY

The view of nomic necessity that first comes to mind upon hearing that laws express relations among universals is that of the Idealists, viz., nomic necessity is but dimly perceived logical necessity (see Ewing (1974)). A neo-Idealist conception of laws has been recently assayed by Christopher Swoyer (1982). For Swoyer, laws express non-contingent relations among properties, or more precisely, relations which are contingent only upon the existence or exemplification of the properties. Hume's ghost stirs. Surely, it quails, for any putative law involving distinct properties, we can imagine a possible world where the properties are exemplified but the relation fails. But on the present proposal, such imaginings are idle, for Hume's move from conceivability to possibility is illicit. You may think that you can dream up a possible world where the electric and magnetic fields do not obey Maxwell's laws, but if Maxwell's equations do indeed express laws, then the world you dreamed up is but a dream and a bad dream at that. The **E** and **B** fields of your dream may be like electric and magnetic fields in various ways, but they are counterfeits, for Maxwell's equations are constitutive of the very nature of electromagnetism.

While I agree that the step from conceivability to possibility must be taken with care, I have no worry that in speculating about a world where magnetic monopoles exist and where, as a result, Maxwell's equations have to be modified, I am changing the meaning or the reference of the term 'electromagnetic field'. Some physicists have seriously proposed that the actual world contains magnetic monopoles. The truth of this proposal does not make me worry that most electro-magnetricians since Maxwell have been referring to nothing at all or else to counterfeit electromagnetic fields. On the other hand, if I am wrong and Swoyer is right, I want to know the error of my ways, and not just in general philosophical terms but in specific cases. But I do not see what evidence would indicate that science had uncovered a relation between **E** and **B** which is truly constitutive of the electromagnetic field vs. a contingent relation which plays a central role in the formulation of the best deductive system. I will return to the epistemological problem for contingent relations among universals in Sec. 12.

There is less agreement on whether Hume succeeded in banishing contingent forms of natural or physical necessity. To decide the matter we need to come to grips with physical necessity at least well enough to know what we are banishing. On the Empiricist conception, it is not coherent to present the metaphysics or semantics of physical necessity by postulating physically possible worlds as a single distinguished proper subset of the logically possible worlds, or to take the relation of nomic accessibility to be a primitive. For the Empiricist, there are no irreducible modal facts. A world W is a world of non-modal facts. Uniquely associated $W \mapsto L_W$ with each such world is a set L_W of non-modal propositions true in W — the laws of W. To mark off the elements of this set we may prefix 'it is physically necessary that', but that prefix is merely an honorific. Accessibility is a defined relation, not a metaphysical given: world W' is nomically accessible from W iff the L_W are all true in W'. There are then myriad subsets of physically possible worlds, each radiating outward from a logically possible world. No one of these collections is more powerful or potent than any other.[13]

We can still ask whether physical necessity so construed can display the trappings of strong necessity in the form of the S_4 and S_5 axioms. Here the S_4 axiom says that physical necessity is robust in that it is transmitted along the relation of nomic accessibility: if W' is nomically accessible from W, then $L_W \subseteq L_{W'}$. The S_5 axiom requires the converse:

if W' is nomically accessible from W, then $L_{W'} \subseteq L_W$. The full S_5 system then requires that $L_{W'} = L_W$. It should be evident that the M-R-L account of how the association $W \mapsto L_W$ is fixed is incompatible with either S_4 or S_5. Suppose, for example, that the laws $L_{W_\oplus}$ of the actual world $W_\oplus$ are the laws of Newtonian mechanics. A world W' containing a single massive particle moving inertially is nomically accessible from $W_\oplus$. But, presumably, the best M-R-L system of W' will not contain Newton's laws but instead an axiom to the effect that all massive particles move inertially; so on the M-R-L version of $W' \mapsto L_{W'}$, $L_{W_\oplus} \nsubseteq L_{W'}$, and $L_{W'} \nsubseteq L_{W_\oplus}$.

Further, from the most basic of the Empiricist constraints, (E0) and (E1), only Confrontational Empiricism is consistent with the full S_5 system; that is, if the S_5 system reigns, the Empiricist laws of two worlds can differ only by being incompatible. The unattractiveness of the ways to affect the association $W \mapsto L_W$ so as to produce Confrontational Empiricism shows why the Empiricist would want to reject even the formalism of strong necessity. There is, for example, Fascist Empiricism: every fact corresponds to a law. Or Imperialist Empiricism: start with a possible world W; choose some set of general propositions true in W and declare them to be the laws L_W of W; take all W's nomologically accessible from W and declare the laws L_W from W to be the laws of W'; choose some W'' outside the first circle and repeat the construction; repeat again and again until all the possible worlds are covered. Any analysis of laws that rejects (E1), while perhaps not being couched in the terminology of strong necessity, will have something of this imperialistic flavor since the putative laws will ride roughshod over the occurrent facts.

Is there any reason to think that physical necessity should follow the dictates of strong S_5 necessity so that, in fine philosophical fashion, the above observations may be turned round and used as an argument against the Empiricist account? At the risk of jousting with straw men, I will note that Tooley's interpretation of the particle example fits with (but does not require) the intuition that nomic accessibility is a symmetric relation so that the laws of $X-Y$ interaction, as evidenced by the regularities of interaction in richer and more sociable worlds, can be brought back to the actual world. I have already said my piece on this example and will say no more. As for the S_4 axiom, I sense that necessitarians — whether they construe necessity in terms of one property yielding or necessitating another or whether they hold a more orthodox modal construal — think that in order for a law to support a counterfactual conditional, the law must not only be true in but must be a law of the world in which the counterfactual situation is imbedded. This is, I believe, a false view of counterfactuals. I will have more to say on counterfactuals in the following section.

12. LAWS AS CONTINGENT RELATIONS AMONG UNIVERSALS

I have to this point neglected the most ambitious attempt to establish the Armstrong-Tooley-Dretske thesis that laws are contingent relations among universals. I will call it the transcendental argument. It has two parts: if laws did satisfy the Empiricist constraints (E1)–(E4), then they would not be able to adequately fill the roles they are supposed to play in supporting subjunctive and counterfactual conditionals, providing explanations, and in grounding induction; and, the argument continues, it is only by adding relations among universals that laws can gain the strength they need to discharge these roles.

Dretske's version of the counterfactual complaint is representative. The complaint is that on the regularity conception of laws, it is a "complete mystery" how laws support counterfactals, for "To be told that all F's are G is not to be told anything that implies that if x were an F, it would be a G" (Dretske (1977), p. 255). True but irrelevant. The real question is whether to be told that it is a *law* that all F's are G is to be told something that implies (or, as I would prefer to say, supports) the conclusion that if this x were an F, it would be G. On behalf of Dretske I will reply: The answer must be negative on, say, the M-R-L version of the regularity account. For then to be told that it is a law that all F's are G is to be told that all F's are G and that this regularity fits neatly with other such regularities to form a strong and simple deductive system. But this just comes down to saying that some regularity, more complex and comprehensive, but no different in kind from all F's are G, holds.

Not so fast! The fact that "All F's are G" is an axiom of the best overall deductive system for this world informs the judgment of similarity we make when comparing other possible worlds to this world. With this information in hand, that "All F's are G" is true in world W_Y but not in W_N is powerful, but not irresistible, persuasion that W_Y is more similar to the actual world than is W_N. Couple this with Lewis's (1973) analysis of counterfactuals and subjunctives and we have a way of seeing how the M-R-L lawhood of "All F's are G" supports the subjunctive "If x were an F it would be a G".

Armstrong's counterfactual complaint (1983, Ch. 5, Sec. 4) is that judgments of comparative similarity are context dependent while the truth and falsity of counterfactuals are not. On the contrary, I think that the logic of counterfactuals is radically context dependent (see van Fraassen (1981)) and that in some contexts we may judge some W_N world to be nearer actuality than any W_Y world and judge it to be false that if x were an F it would be a G. But if one does not want such results, then it can be added as a constraint on comparative similarity that laws always have an overriding priority in assessing similarity. If laws are contingent — whether on occurrent facts or on non-occurrent facts about the relations of universals — then I do not see that there is any other alternative;[14] unless, that is, one is prepared to offer a wholly different analysis of counterfactuals.

I turn now to explanation. Hardbitten Empiricists are apt to disparage the notion of explanation. The Quine of "Necessary Truth" allows that "in natural necessity, or our attribution of it, I see only Hume's regularities, culminating here and there in what passes for an explanatory train or promise of it" (1976, p. 76). The Wittgenstein of the *Tractatus* was more straightforward: "The whole modern conception of the world is founded on the illusion that so-called laws of nature are explanations of natural phenomena" (1961, 6.371). Empiricists would do better, I think, to accommodate the notion of scientific explanation. It is the universalist's contention that no such accommodation is possible. Thus, Dretske writes:

The fact that every F is a G fails to explain why *any F* is a G . . . The fact that all men are mortal does not explain why you or I are mortal; it *says* (in the sense of *implies*) that we are mortal, but it does not even suggest *why* this might be so . . . Subsuming an instance under a universal generalization has as much explanatory power as deriving Q from $P \cdot Q$. None. (1977, p. 262)

Professor Armstrong's complaint is similar:

All F's are G's is a complex state of affairs which is in part *constituted* by the fact that all observed F's are G's. 'All F's are G's' can even be rewritten as 'All observed F's are G's and all unobserved F's are G's'. As a result, trying to explain why all observed F's are G's by postulating that all F's are G's is a case of trying to explain something by appealing to a state of affairs part of which is the thing to be explained. (1983, p. 40)

The remedy for this situation is supposed to be a linkage of necessitation between the universals F and G; in Armstrong's notation, $N(F, G)$, read "F-ness necessitates G-ness". But the explanatory force of such a linkage has got to derive not from the strength of the necessitation but from its quality; as Armstrong warns, even the strongest reasons for believing *that* something is or must be the case need not explain *why* it is the case. What then is the quality of $N(F, G)$ which confers explanatory power? Dretske's answer is that $N(F, G)$ (or in his notation, $F \rightarrow G$) explains why this F is a G because it means that "F-ness is linked to G-ness" in the sense that "the one property yields or generates the other" (1977, p. 264). Armstrong's answer is that $N(F, G)$ explains why all (observed) F's are G's because it unifies the instances of the regularity (1983, Ch. 6).

The Empiricist has a ready response to these concerns. What the universalists seek to achieve through ontological ascent, the Empiricist achieves by ascent of explanatory level. Unification of observational regularities is achieved by passing to higher level laws while evolutionary accounts of how one set of properties yields or generates another set are to be found in the dynamical laws of physics. The truth of Kepler's law, "For every planet, the radius vector from the sun to the planet sweeps out equal areas in equal times," may not explain why this phenomenon is so for the earth and for all observed planets. But Newton's laws of motion coupled with his law of gravitation do explain Kepler's generalization and, thus, why particular instances conform to it. The explanatory malaise feigned by the universalists is diagnosed not as a symptom of a defect in Empiricist explanations but as a result of the artificiality of philosophical discussions where the 'laws' discussed are of the "All ravens are black" variety.

The universalists may reply that at each ascending level the explanatory malaise arises again. We will, for example, want to know why the sun and the earth attract each other with a force inversely proportional to the square of the distance between them, and Newton's force law does nothing to help us here since it simply asserts that this is so for any pair of massive bodies. The Empiricist must concede that his brand of empiricism does not provide for ultimate explanations in the you-can't-ask-for-anything-more sense. But then neither does the universals conception. I may agree that being massive yields or generates gravitational attraction, but this does not block my request for a fuller and deeper understanding of the how and why of gravitation if, as

Armstrong and Dretske would have it, the Idealists are wrong and the relation between massiveness and gravitational attraction is contingent. It took more than two centuries before the how and why were revealed by Einstein. But his revelations came not in the form of a discovery of a bonding of universals missed by Newton but in the form of a new level of explanation in terms of space-time warps.

I turn now to a closer look at the nature and status of $N(F, G)$. $N(F, G)$ is supposed to be contingent and more so than Swoyer's thin sense that it depends on the existence or exemplification of the universals F and G. But $N(F, G)$ is not contingent on occurrent facts — (E1) is violated; rather it is contingent on another category of facts which transcend the occurrent. How then do we have epistemological access to $N(F, G)$? If W_1 and W_2 share all occurrent facts, they are, by Empiricist lights, the same world. For those who say otherwise the Empiricist will crank up the unknowability argument, rehearsed briefly in Sec. 3, in order to show that W_1 and W_2 are epistemologically indistinguishable; so if $N(F, G)$ holds in one but not the other of these worlds, we could never know which was which. This indistinguishability claim can be attacked in two ways. First, one could challenge the Empiricist premise that only occurrent facts can be known directly or non-inferentially and try to show how direct knowledge of $N(F, G)$ can be obtained. Second, one could challenge the second Empiricist premise that what is underdetermined by everything which is in principle directly knowable is unknowable in principle and try to show how inferential knowledge of $N(F, G)$ is sustained. Armstrong takes the first route and Tooley the second.

At the end of Vol. II of *Universals and Scientific Realism* Armstrong suggests that we have non-inferential knowledge of nomic necessitation via direct perceptual awareness of instances of causal connections (1978, Vol. II, pp. 162ff). This corresponds to knowledge of the intermediate link in the chain

$$N(F, G) \Rightarrow (x)N(Fx, Gx) \Rightarrow (x)(Fx \supset Gx)$$

where the arrow is an entailment relation[15] and $N(Fx, Gx)$ means that x's being F necessitates its being G. Strictly speaking, what we have is direct knowledge that one event necessitates another, which knowledge becomes upon reflection the knowledge that there exist universals F and G and particular x such that x's being F necessitates its being G. I wish Armstrong had used *What is a Law of Nature?* to elaborate further on his earlier remarks. In the absence of further elaboration I waive the well-worn Humean objection that upon reflection our knowledge of causal sequence becomes knowledge of constant conjunction and/or felt determination. But I do not waive my conviction that if there is nomic necessitation, its ultimate springs are most likely hidden from our view. The ultimate laws of nature, whatever they may be, will most likely involve universals whose instancings correspond to states of affairs which are not directly observable and which are thus knowable only inferentially.

The subject of unobservability suggests an analogy that may be helpful to the universalist in impeaching the second premise of the Empiricist unknowability argument. Strict Empiricists have sometimes sided against a realistic interpretation of scientific theories on the grounds that theories are underdetermined by everything that is in principle knowable by direct observation. Scientific realists respond that such underdetermination is not fatal because we can have general reasons for believing in the existence of unobservable theoretical entities and specific reasons for believing one observationally equivalent theory over another. The suggestion then is that the universalist try to show that he can parallel the scientific realist's response and in this way demonstrate that realism with respect to relations among universals is no worse off than scientific realism in general.

Some forms of the scientific realist responses do not appear to lend themselves to this piggyback strategem. Unless Armstrong is correct about our having direct perceptual awareness of instances of nomic necessitation, the universalists cannot avail themselves of the slippery slope response; viz., there is a blurred and shifting line between what is and is not observable, with yesterday's unobservable becoming tomorrow's observable. Nor can the universalist latch onto the goals of science response; viz., in order to achieve, say, deterministic laws or the linking together and systematization of observational regularities, it is necessary to ascend to the theoretical level. Examples due to van Fraassen (1980) and Rynasiewicz (1981, 1983) show quite conclusively that the observational content of a theory T cannot be identified with the set of observational sentences logically implied by T. A better construction might go something like this. Start with the models of the axioms of T; then take their observational reducts;[16] then restrict the domains of the reducts to objects which are directly observable. The resulting set of models corresponds to the class of observational states of affairs allowed by T. But most likely this class will not be an elementary class in even the wider sense; that is, it will not be the set of models satisfying some countable set of sentences (see Rynasiewicz, 1981), so even minimal systematization is not possible at this level. Of course, the strict Empiricist will rejoin that the systematization afforded by T is a pragmatic virtue, providing a reason to use T but no reason to believe that its theoretical assertions are true. But this is a dispute I do not wish to enter here.

Some scientific realists have held that prior probability considerations can supply the grounds for favoring one theory over another even when the theories are observationally equivalent. This strategy can by piggybacked by the universalists, but I wonder whether Bayesianism is the sort of piggy they want to back. The objectivist conception of prior probabilities remains nothing more than a collection of vague promissory notes. And the more popular subjective degree of belief conception seems to cut little philosophical ice for the case under discussion. Subjective degrees of belief can be assigned to hypotheses about relations among universals, but then they can also be assigned to hypotheses about anything you like — devils, angels, vital forces as well as electrons. Numerical representations of opinions may be helpful for certain purposes, but one expects more than mere representation of opinions from an account of the testing and confirmation of scientific theories.

More is to be found in Glymour's *Theory and Evidence* (1980). Glymour offers an objectivist account of qualitative confirmation which overcomes some of the more egregious flaws of the hypothetico-deductive view. His approach is essentially an extension of Hempel's idea that hypotheses are confirmed by deducing instances of them from evidence statements. Glymour's ingenious addition to this idea is a 'bootstrapping' operation by which instances of theoretical hypotheses are deduced from observational evidence with the assistance of auxiliary hypotheses drawn from the theory being tested.[17] I doubt that there is help to be found here for the universalists. If 'instances' of $N(F, G)$ are instances of $(x)(Fx \supset Gx)$, then Glymour's account can help to show how confirming instances are obtained from observational evidence when 'F' and 'G' denote properties that are not directly observable. But then what is being bootstrap confirmed is not $N(F, G)$ but its extensional counterpart. On the other hand, if 'instances' of $N(F, G)$ are instances of nomic necessitation, e.g., $N(F, G)$ (a's being

F, *a*'s being *G*),[18] then even when '*F*' and '*G*' denote directly observable properties, I do not see how Glymour's method can be used to confirm a theory with axioms like *N*(*F*, *G*), at least not if, *pace* Armstrong, observational evidence comes in the Humean form *Fa* & *Ga*.

In sum, I can find no reason to share Tooley's optimistic conclusion that whatever account can be given for the grounds for accepting scientific theories in general will serve as well as an account of the grounds for accepting *N*(*F*, *G*).

I can already hear the reply of Profs. Armstrong and Dretske; viz. it is not a matter of giving grounds for accepting *N*(*F*, *G*); rather, relations among universals are presuppositions for induction and confirmation. But I contend that relations such as *N*(*F*, *G*) are not presuppositions in the sense of conditions without which the wheels of confirmation would not turn.[19] If it is said that such relations are needed to make the machinery of confirmation intelligible, then we have reached an impasse. I can no more accept this standard of intelligibility than the ones set up by the Idealists and Rationalists.

In closing I want to make it plain that I do not suffer from one of those strange afflictions that make some of my colleagues hanker after desert landscapes or pant after particulars. I am fully convinced that universals occupy an important place in our ontology. And I reject Ramsey's gibe: "But may there not be something which might be called real connections of universals? I cannot deny it for I can understand nothing by such a phrase ..." (1978, p. 148). I *can* understand something by such a phrase, but my understanding of the use to which Armstrong, Dretske, and Tooley want to put it is incompatible with my understanding of empiricism.

13. CONCLUSION

What is missing in this chapter, and in most of the philosophical literature reviewed here, is the texture and feel of real-life laws. That is the sacrifice we made in attempting to abstract features common to all natural laws. The reader will have to judge for herself whether the results have justified the sacrifice.

I have attempted to obey Braithwaite's injunction to remain within the ambit of the constant conjunction view when giving the rationale for the distinction between uniformities due to natural laws and those which are merely cosmic accidents. The Attitudinal theorists despair of being able to satisfy this injunction, and in their despair they relocate the source of the distinction from the uniformities to our attitudes towards them. The Necessitarians and Universalists share this sense of despair, but rather than resort to human attitudes they appeal, in the former case, to irreducible *de re* modalities, and in the latter case, to contingent relations among universals. I argued on empiricist grounds that Necessitarian and Universals views are unacceptable. And I tried to show that while there is no easy path to the satisfaction of Braithwaite's injunction, neither is there sufficient reason to submit to the despair of the Attitudinalists.

A regularity analysis denies to laws of nature various forms of necessity that some philosophers claim for them. This denial in turn removes from determinism some of the sting Libertarians have felt; whether enough of the string is drawn to resolve the determinism-free will problem is an issue to be discussed in Ch. XII.

In previous chapters we have seen reasons for rejecting the notion that determinism is an *a priori* truth or an indispensible presupposition of scientific enquiry, but we have also seen that the force of determinism is not captured by saying that it is merely a high level empirical claim or a useful methodological guideline. The present chapter confirms the special and peculiar status of determinism; for while it is not essential to laws, it can and often does promote both strength and simplicity, the combination of which we took (following Mill) to be the essence of lawhood.

In Ch. III I suggested that determinism can be used as a probe for exploring the concepts of physical possibility and necessity. There is an obvious and innocuous sense in which this suggestion can be taken: namely, posit determinism and then see what presuppositions are needed to make it work; these presuppositions then become candidates for inscription on the list of natural laws. This suggestion involves no circularity if the standards for judging candidacy do not themselves presuppose determinism. And here we confront a difficulty: to the extent that the measures of strength, simplicity, coherence, etc. used in the M-R-L account of laws are not biased for or against determinism, they are not precise enough to cleanly decide some of the tough questions about determinism for Newtonian and relativistic physics. Is the deductive system of Newtonian mechanics with the boundary conditions at infinity needed to secure Laplacian determinism better than the system without these boundary conditions? Ditto for the boundary conditions needed to make the classical heat equation deterministic. Ditto for the entropy conditions needed to make the shock wave equation deterministic. Ditto for the prohibition against tachyons. Ditto for the conditions on the null cone structure needed to make general relativistic worlds deterministic (see Ch. X). My own answers are (in order): No; No; Yes; Don't know; No; with the overall tally going against determinism. But in every case I have to admit that my judgment is unstable and has more the feel of an esthetic judgment than a scientific judgment. And as in matters of esthetics, others give a different series of judgments, with their tallies often more in favor of determinism than mine. That the doctrine of Laplacian determinism has no firm truth value for Newtonian and classical relativistic physics is a conclusion some will find intolerable. Intolerable or not, the ambiguity is one we have to live with, at least until someone can fashion the tools to resolve it.

NOTES

[1] These are the versions of the constant conjunction and felt determination definitions Hume gives in the *Treatise*. The definitions are repeated with some significant changes in the *Enquiry*.

[2] This counterfactual definition does not appear in the first edition of the *Enquiry*.

[3] Here I am following Suchting (1974).

[4] See especially Secs. 13 and 15 of Bk. I of the *Treatise*.

[5] For a more detailed discussion of this point, see Suchting (1974) and Armstrong (1983).

[6] For an attempt to fill in some of the details, see Reichenbach (1954).

[7] Unlikely but not impossible since the net impressed force acting on a particle can be zero even when other particles are present. But the point is that we do not want the lawfulness of Newton's First Law to turn on such a happenstance.

[8] This last move would yield the stronger version of (E1); namely, if W_1 and W_2 agree on all occurrent facts, then they are the same world.

[9] See, for example, Braithwaite (1960), Berofsky (1968) and Tondl (1973); see Suchting (1974) for a critical discussion.

[10] It must be admitted that from this perspective Mill's rather labored treatment of Reid's famous day-night example is anomolous; see Mill (1904, pp. 244—247).

[11] The only finite frequentists I can cite are Russell (1948) and Sklar (1973).

[12] For a discussion of frequency and propensity theories and references to the literature, see Ch. VIII below.

[13] The truth value semantics for this type of approach have been worked out by Dunn (1973).

[14] This is one of the considerations that persuade Swoyer (1982) that laws are contingent only upon the exemplification of the universals.

[15] Hochberg (1981) has questioned the nature of this entailment relation. Armstrong's answer (1983, Ch. 6) seems satisfactory to me. What remains to be worked out is the formal semantics of the entailment relation; whether this can be done consistently with the constraints Armstrong places on universals remains to be seen.

[16] Roughly, just lop off all the terms which do not correspond to directly observable properties and relations.

[17] For Glymour the basic confirmation relation is three-place: evidence E confirms hypothesis H relative to theory T. We can say that E confirms T iff there is an axiomatization of T such that E confirms each axiom A relative to T. Glymour originally allowed the use of the hypothesis H itself as an auxiliary in deducing instances of H from E. But in later versions this feature has been dropped; see the articles by van Fraassen and Edidin in Earman (1983).

[18] Read: "a's being F necessitates a's being G in virtue of the universals F and G."

[19] Dretske (1977) says that lawfulness must be assumed for a general hypothesis H if examined instances which conform to H are to raise the probability that unexamined instances also conform to H. This I deny. If $Pr(H) \neq 0$, then it is a theorem of probability that as the number of examined instances conforming to H approaches infinity, the probability that any number of unexamined instances also conform to H approaches 1. Perhaps it may be claimed that it is unreasonable to set $Pr(H) \neq 0$ unless H is backed up by the appropriate relations among universals. This I also deny.

SUGGESTED READINGS FOR CHAPTER V

Chs. 4 ("Of Laws of Nature") and 5 ("Of the Law of Universal Causation") of Bk. III of Mill's (1904) *System of Logic* are the source of the modern regularity account of laws. Suchting's (1974) article "Regularity and Law" and the first part of Armstrong's (1983) book *What Is a Law of Nature?* detail the reasons why a growing number of philosophers have become disenchanted with the regularity analysis. The view that laws of nature are relations among universals is set out in the second half of Armstrong's book and in Dretske's (1977) "Laws of Nature" and Tooley's (1977) "The Nature of Laws." Skyrms' (1980) *Causal Necessity* offers a resiliency analysis of laws that can be traced back to Mill. Ayer's (1956) "What Is a Law of Nature?" and Goodman's (1955) *Fact, Fiction and Forecast* defend a felt determination definition of laws. Necessitarian accounts of laws are to be found in Ewing's (1974) *Idealism* and Kneale's (1949) *Probability and Induction.*

CHAPTER VI

DETERMINISM, MECHANISM, AND EFFECTIVE COMPUTABILITY

"Garbage in . . . garbage out"
"Yeah but is it computable garbage?"
(Graffiti from wall of Men's Room,
Experimental Engineering Bldg.,
University of Minnesota)

The examples of determinism studied in previous chapters should make it clear that determinism does not entail mechanism in the crude sense that determinism necessarily works by means of a mechanical contrivance composed of gears, levers, and pulleys. But it remains open that determinism involves mechanism in the more abstract sense that it works according to mechanical rules, whether or not these rules are embodied in mechanical devices. In the converse direction we can wonder whether mechanistic rules are necessarily deterministic. To make such questions amenable to discussion we need a model of mechanism and a codification of the rules by which the model works. I will take as the starting paradigm of mechanism the device which increasingly and irresistibly colors modern life — the digital computer. To understand the gist of operation of these devices it is best not to get too abstract too quickly, but to begin with the minimal embodiment described by Alan Turing in 1937.

1. TURING MACHINES

The inputs and outputs to a Turing machine are recorded on an infinite paper tape which is divided into squares. In each square one of three symbols, '0', '1', or 'B', appears. In its pristine state, before input, the tape is completely blank ('B' printed in each square). The machine 'scans' one square at a time and performs one of the following basic operations: it erases the symbol in the square it is currently scanning and prints one of the other symbols; it shifts one square to the left; it shifts one square to the right; or it puts up a flag and halts. For sake of definiteness, we can suppose that one basic operation is performed per second. The sequence of operations is governed by a finite list of

111

deterministic rules. The guts of the machine, as distinguished from the tape, are at any instant in one of a fixed finite set of states, $s_1, s_2, \ldots,$ s_N, one of which is the starting state s^* and another of which is the halting state s^{**}. The rules of performance have the following form: if at t_i the internal state is ___ and the symbol on the square being scanned is ___, then perform the basic operation ___ and shift into state ___ at t_{i+1}. Determinism here simply means that no two rules have the same filling for the first two blanks but a different filling in either of the last two blanks, and that there is a rule to cover each possible combination of initial internal state and tape symbol. It is also understood that the rules are time translation invariant, i.e., they are independent of the index i on the time t_i. An input to the machine will be a code $\bar{m}$ for a natural number m with '0' representing 0 and a string of m consecutive '1's representing a positive m. By convention, the input code is flanked on both sides by 'B's with the first blank to the right of the code being the square scanned when the machine is in the start state. If for given input the machine does not halt, then by definition there is no corresponding output. But if for given input the machine does halt, then the corresponding output is defined as the tape code when s^{**} is reached, and it is arranged that the output code is flanked by 'B's with the machine resting on the first 'B' to the left of the output (see Fig. VI.1).

Fig. VI.1

With any such machine we can associate a partial function $f: \mathbb{N} \to \mathbb{N}$: for $m \in \mathbb{N}$, if the machine does not halt for input $\bar{m}$, then $m \notin \text{dom}(f)$; if the machine does halt for input $\bar{m}$ giving output $\bar{n}$, then $f(m) = n$. A (partial) function of the natural numbers will be said to be *Turing computable* just in case it is associated with some Turing machine.

It is intuitively compelling that any such function should count as being effectively, mechanically computable, at least in principle. What is not so clear is the converse. Part of the worry here can be removed by proving that we do not get a larger class of computable functions by enlarging the alphabet, or by allowing the machine to scan more than one square at a time, or by using two tapes instead of one, etc. But such results do nothing to assuage the worry that a mechanical device operating in a very different fashion could compute a Turing uncomputable function. For example, we might start with the Turing hardware but operate it by non-deterministic rules which, for given internal state and state of the square being scanned, allow for a (finite) number of choices consisting of the next basic operation and the next internal state. Of course, such a non-deterministic Turing machine cannot be directly regarded as a function computing device if, for some input, it halts for some but not all subsequent histories or else gives different outputs for different halting histories. Nevertheless, we will say that f is non-deterministically computable just in case there is a (possibly) non-deterministic Turing machine such that for each $m \in \mathrm{dom}(f)$, the machine halts in some allowable history following the input $\bar{m}$, giving the output $\bar{n}$ where $n = f(m)$, and for $m \notin \mathrm{dom}(f)$, the machine does not halt in any allowable history following the input $\bar{m}$. Intuitively, such a function should count as being effectively computable; for we can effectively generate sequences of selections from a finite number of choices, and combining such a sequence with a non-deterministic Turing machine gives unambiguous instructions for computing a (partial) function. So, evidently, determinism is not essential to this form of mechanism. But any function computable by a non-deterministic Turing machine can be computed by a standard Turing machine. This can be seen by, first, using a three tape Turing machine which employs one tape to hold the input, a second to generate a sequence of choices, and a third to reproduce the results of one of its non-deterministic cousins for the generated choices, and by, secondly, appealing to the fact that a standard Turing machine can do everything a multi-tape version can do. (In a sense, computing power may be gained by going to non-deterministic machines, for by making clever or lucky choices non-deterministic Turing machines may be able to accomplish tasks in a smaller number of steps than their deterministic brethren. The following is, apparently, an open question: If $f(x)$ is computable by a non-deterministic Turing machine in a time $t(x)$ which is a polynomial in x, is it also computable in polynomial time on a deterministic machine?)

One can still worry that some altogether different device, not resembling either a deterministic or non-deterministic Turing machine in hardware or software, could compute a function not computable by either type of Turing machine. Such doubts can never be entirely banished, but they are mitigated by the remarkable convergence of a number of independent lines of investigation. For example, Kleene offered a different definition of effectively computable functions of $\mathbb{N}$ using the concept of recursiveness; but his definition picks out exactly the same class of computable functions as does Turing's. And other definitions by Church, Markov, and Post and others also prove to be extensionally equivalent to Turing's.[1]

Church's thesis (as it is called) that effective, mechanical computability for functions of $\mathbb{N}$ is to be identified with Turing computability, is now accorded such faith that it is an acceptable mode of informal argumentation to conclude Turing computability or recursiveness from the existence of an informal algorithm.[2] The trick is to recognize when an informal procedure corresponds to a genuine algorithm.

2. DETERMINISM AND EFFECTIVE COMPUTABILITY: FIRST TRY

The starting question as to whether determinism implies mechanism can now be reformulated as a series of questions about effective computability:

If the laws L are Laplacian deterministic, does it follow that there is an effective procedure for generating the solutions of initial value problems? Will the (unique) solution of any given initial value problem be an effectively computable function? Will any solutions determined by effectively computable initial data be effectively computable? Will effectively computable initial data always be transformed into data which, at each future instant, will also be effectively computable?

To facilitate thinking about these questions, consider a deterministic discrete state system that operates in discrete time. At each instant, $t = 0, \pm 1, \pm 2, \ldots$, the state of the system is given by specifying a non-negative integer ('occupation number') for each of an infinite number of slots (e.g., number of balls in an urn, number of atoms at a given energy level). Thus, a history of the system is a function ϕ: $\mathbb{N} \times \mathbb{Z} \to \mathbb{N}$, where $\phi(m, n)$ is the occupation number of slot m at time n. The reader can amuse herself by constructing examples where the allowed histories form a deterministic set but the questions posed above have negative answers. Try, for example, to design deterministic transition laws so that for some allowed history ϕ, $\phi_0(\cdot) \equiv \phi(\cdot, 0)$ is Turing computable (i.e., the occupation numbers at time $t = 0$ are effectively enumerable) but for some $n > 0$, $\phi_n(\cdot) \equiv \phi(\cdot, n)$ is not

Turing computable (i.e., the occupation numbers at $t = n$ are not effectively enumerable).

However, there is little payoff to be gained in pursuing such examples unless they can be brought to bear on realistic physical systems. And here we meet a conundrum: all of the examples we have studied from physics involve functions of the real numbers, but, so far, we have given to characterization of effective computability for such objects. A construction by Grzegorczyk promises to fill the gap.

But before turning to functions of the reals, it is worth noting that the discrete state machine contemplated above affords another means of characterizing effective computability for functions of the integers that is extensionally equivalent to Turing's but conceptually more appealing. An *unlimited register machine* (URM)[3] consists of an infinite number of registers R_1, R_2, ... each of which holds a natural number r_n. A program for operating this machine consists of a finite list of instructions of four basic types. First, a zero instruction changes the contents of a designated R_n to 0 while leaving the other registers unaffected. Second, a successor instruction adds 1 to the contents of a designated R_n while leaving the others unaffected. Third, a transfer instruction interchanges the contents of two designated registers R_n and R_m again leaving the others unaffected. And finally, a jump instruction compares the contents of R_n and R_m and orders the machine to proceed to intruction number q or else to the next instruction according as $r_n = r_m$ or not. A (partial) function $f: \mathbb{N}^n \to \mathbb{N}$ is said to be URM computable if there is an URM which computes it in the following sense: if $(a_1, a_2, \ldots, a_n) \notin \text{dom}(f)$, then the machine does not halt when the initial contents of the registers are $a_1, a_2, \ldots, a_n, 0, 0, \ldots$; but if $(a_1, a_2, \ldots, a_n) \in \text{dom}(f)$, then with the initial contents $a_1, a_2, \ldots, a_n, 0, 0, \ldots$ the machine does halt with $b = f(a_1, a_2, \ldots, a_n)$ in R_1. (To make the URM behave as a deterministic system in our sense, with the contents of the registers at any time determining the contents at any later time, we would need to add an initial register R_0 to hold the number of the next instruction to be executed and also add to the programming instructions a rule to modify r_0 in the appropriate way. I leave it to the reader to supply the details.)

Combining the remarks from the first part of the section with the URM characterization of computability reawakens worries about Church's thesis. There are innumerable numbers of deterministic ways to run the register machine that outstrip any standard URM. Why can't some of these ways be used to compute a Turing uncomputable function? Such questions will be held in abeyance until Sec. 7.

3. GRZEGORCZYK COMPUTABILITY

The strategy is to move along the well-charted path from the integers through the rationals to the reals, effectivizing definitions as we go. Thus, a sequence of rational numbers $\{x_n\}$ is said to be effectively computable just in case there are three Turing computable functions a, b, c from $\mathbb{N}$ to $\mathbb{N}$ such that $x_n = (-1)^{c(n)} a(n)/b(n)$. A real number r is said to be effectively computable if there is an effectively computable sequence $\{x_n\}$ of rationals which converges effectively to r, i.e., there is an effectively computable function d from $\mathbb{N}$ to $\mathbb{N}$ such that $|r - x_n| < 1/2^m$ whenever $n \geq d(m)$. Taking $\{x'_n\} = \{x_{d(n)}\}$, it follows that $|r - x'_n| < 1/2^n$ for all n. Continuing in this vein, a sequence $\{r_n\}$ of reals is said to be effectively computable if there is a computable double sequence $\{x_{kn}\}$ of rationals such that $|r_k - x_{kn}| < 1/2^n$ for all k and n.

It remains to say what an effectively computable function of the reals is. Grzegorczyk's (1955) concept of effective computability for a function f of the reals was originally stated in terms of recursive functionals, but this definition is not at all easy to apply to concrete examples in analysis. Grzegorczyk (1957) showed that the original definition is equivalent to several others, including the following which is the one most often used by analysts:

(i) f is sequentially computable: for each effectively computable sequence $\{r_n\}$ of reals, $\{f(r_n)\}$ is also effectively computable, and

(ii) f is effectively uniformly continuous on rational intervals: if $\{x_n\}$ is an effective enumeration of the rationals without repetitions, then there is a three place Turing computable function l such that $|f(r) - f(r')| < 1/2^k$ whenever $x_m < r$, $r' < x_n$ and $|r - r'| < 1/l(m, n, k)$ for all $r, r' \in \mathbb{R}$ and all $m, n, k \in \mathbb{N}$.[4]

Yet another equivalent definition in terms of an effective polynomial approximation of f is given by Pour-El and Caldwell (1975).

Provisionally accepting Grzegorczyk's definition, we will go on to use it to answer questions about the relation between determinism and effective computability in physics.

4. DETERMINISM AND EFFECTIVE COMPUTABILITY: ORDINARY DIFFERENTIAL EQUATIONS

Consider the first order ordinary differential equation

(VI.1) $\dot{\phi}(t) = F(t, \phi(t))$

subject to the initial condition

(VI.2) $\phi(0) = \phi_0$

Uniqueness for the initial value problem is not guaranteed if we merely required continuity of F. Suppose in addition we demand that F satisfy a Lifshitz condition on a rectangle about the origin. (Recall that this means that there are constants α, β, and K such that $|F(x, y) - F(x, y')| \leqslant K|y - y'|$ for all (x, y) in the rectangle $|x| \leqslant \alpha$ and $|y - \phi_0| \leqslant \beta$. K is called the Lifshitz constant.) Then local existence and uniqueness theorems for the initial value problem are forthcoming.

Moreover, the existence theorem actually provides an effective procedure for cranking out a sequence of approximations converging to the (unique) solution. The 0th approximation is just the initial value ϕ_0. The next approximation is

(VI.3) $\phi_1(t) = \phi_0 + \displaystyle\int_0^t F(\xi, \phi_0)\, \mathrm{d}\xi$

and in general

(VI.4) $\phi_k(t) = \phi_0 + \displaystyle\int_0^t F(\xi, \phi_{k-1}(\xi))\, \mathrm{d}\xi, k = 1, 2, 3, \ldots$

The convergence of this sequence is uniform, and it is also effective since given α, β, K and the bound $M \geqslant |F(x, y)|$ on the rectangle, we can effectively compute how many times the crank needs to be turned to come within the desired approximation of the solution. Further, if F is an effectively computable function and the initial value ϕ_0 is an effectively computable number, then the approximating functions ϕ_k are also effectively computable since plugging a computable number in a computable function, composing computable functions, and integration are all computability preserving operations. The upshot is that if F and the initial data are computable, then so is the (unique) solution; for if $\phi_k \rightarrow \phi$ uniformly and effectively, then ϕ is Grzegorczyk computable if the ϕ_k are.

Of course, we know by cardinality considerations that most of the solutions of (VI.1)–(VI.2) will not be Grzegorczyk computable even when F is; for there are an uncountable number of solutions but only a countable number of computable functions. But it is remarkable that the solutions picked out by computable initial data are computable and that there is an effective procedure for generating them.

When the conditions needed to prove uniqueness are relaxed, then even though F is computable, the non-unique solutions need not be; in fact, there are cases where none of the solutions are computable (see Pour-El and Richards (1979)).

Thus, there is a strong and deep connection between determinism and effective computability for first order ordinary differential equations. Higher order ordinary differential equations can sometimes be reduced to a system of first order equations, in which case the connection between determinism and computability carries over. For partial differential equations the story is both more complicated and more interesting.

5. DETERMINISM AND EFFECTIVE COMPUTABILITY: PARTIAL DIFFERENTIAL EQUATIONS

For the non-linear shock wave equation (III.7) we saw that initial data $u(x, 0) = u_0(x)$ determines a unique weak solution $u(x, t)$, $t > 0$, if entropy conditions are imposed. We can choose $u_0(x)$ to be very smooth and Grzegorczyk computable. But in general the computability of the initial data is not preserved since $u_c(x) = u(x, c)$, $0 < c =$ constant, may not be continuous and, therefore, not Grzegorczyk computable.

For the classical heat equation (III.4) we saw that Laplacian determinism in the future direction holds if supplementary boundary conditions at infinity are imposed. We also saw that in the unique future solution the smoothness of the initial data does not degrade — just the opposite, any roughness in the initial temperature distribution disappears after even so short a time — so that a breakdown in computability cannot occur because of a loss of continuity. And in fact, a computable initial temperature distribution determines a computable solution (see Pour-El and Richards (1983)).

The discussion of the relativistic wave equation (IV.1) requires that we respect the separation already made between the case of one-dimensional space and the case of higher dimensions. In the former case the form of the solution

$$(VI.5) \quad u(x, t) = \tfrac{1}{2}[f(x + t) + f(x - t)] + \int_{x-t}^{x+t} g(\xi)\, d\xi$$

shows that if the initial functions $f(x) = u(x, 0)$ and $g(x) = \partial u(x, 0)/\partial x$ are computable, then so is the solution $u(x, t)$. In higher dimensions we saw that C^1 initial data can degrade to C^0 but not below, so computability for solutions determined by differentiable initial data will not break down for the sorts of reasons that applied to the shock wave case. Nevertheless, Pour-El and Richards (1981) have constructed examples for the three-dimensional case where $g(x^1, x^2, x^3) = 0$, $f(x^1, x^2, x^3)$ is C^1 and computable, but the unique C^0 weak solution is not Grzegorczyk computable. Further, they give an example where $f(x^1, x^2, x^3)$ is computable and therefore continuous but the corresponding solution $u_1(x^1, x^2, x^3) = u(x^1, x^2, x^3, 1)$ at $t = 1$ is continuous but not computable.

6. EXTENDED COMPUTABILITY

On Grzegorczyk's definition, discontinuity automatically brands a function of the reals as being non-computable. This is counterintuitive,[5] as indicated by the simplest example of a discontinuous function, a step function such as $s(x) = c$ for $x < x_0$, d for $x \geqslant x_0$. If the jump point x_0 and the jump values c and d are computable numbers, then we would like to be able to count s as being computable. This can be done while maintaining contact with Grzegorczyk's approach, for we can say that the step function s is computable because it can be effectively approximated by Grzegorczyk computable functions, at least if we are willing to measure the degree of approximation in a sufficiently liberal way. For simplicity, restrict attention to functions defined on a compact interval $[r_1, r_2] \in \mathbb{R}$. The $L^p[r_1, r_2]$ norm for such a function is defined as

$$\|f\|_p \equiv \left[\int_{r_1}^{r_2} |f(\xi)|^p\, d\xi \right]^{1/p}$$

It is easy to see that taking the effective closure of Grzegorczyk computable functions in, say, L^1, captures our step functions. It turns out the choice of p does not matter much: for any $1 \leqslant p < +\infty$ the step functions captured in this way are exactly the ones with computable jump point and jump values (Pour-El and Richards (1983)). In general, we can say that f is computable in the norm $\| \ \|$ if there is a sequence of Grzegorczyk computable functions $g_1, g_2, g_3, \ldots$, such that $\|f - g_k\|$ converges effectively to 0 as $k \to \infty$. If $\| \ \|$ is the usual sup norm then we collapse back to the original class of Grzegorczyk functions.

We now have to rework the cases where determinism did not preserve Grzegorczyk computability and ask whether computability in the extended sense for some norm appropriate to the problem is preserved. The qualifier 'appropriate' introduces an annoying vagueness, but in various concrete cases bounds on appropriateness are usually evident.

For the relativistic wave equation what was true for Grzegorczyk computability continues to be true for the extended concept of L^p computability: in three spatial dimensions the wave equation does not preserve computability in the L^p norm ($p < +\infty$). However, in the energy norm, computable initial functions determine a computable solution (see Pour-El and Richards (1983)).

For the shock wave equation we can restrict attention to cases where $u(x, 0) = u_0(x)$ has compact support; then, $u_c(x) = u(x, c)$, $c > 0$, also has compact support. The appropriate norm here seem to be L^1, and for this choice computability in the extended sense need not be preserved since discontinuities in $u_c(x)$ need not occur at computable points or with computable jump values.

7. GENERALIZED COMPUTABILITY

Turing machines, URMs, and the other devices commonly used to characterize effective computability are very special examples of deterministic systems. It is therefore natural to wonder whether more general deterministic systems can be used to 'effectively compute' functions which are not Turing-Grzegorczyk computable. The scare quotes are an acknowledgment of the danger that too hasty a generalization may lead to triviality. If we allow the initial state of an URM to contain too much information, any total function f of $\mathbb{N}$ becomes 'computable'. Take the initial contents of the registers to be a, $f(1)$, $f(2)$, ... and take the program instructions to transfer the contents of R_1 and R_{a+1} and halt. A similar trivialization occurs if the program is allowed to contain an infinite list of instructions. And beyond these trivial trivializations, other more interesting ones lurk. While being aware of the trivialization danger, we should not allow it to prevent us from exploring non-Turing notions of computability.

For functions of the reals more than idle curosity motivates the exploration. It seems more natural to try to characterize computability for these functions directly in terms of analogue computers which are designed to handle continuous variables than in terms of computations on rational approximations to reals. It remains to be seen how analogue computability of functions of the reals is related to Grzegorczyk computability.

A paradigm example of a general purpose analogue computer is the differential analyzer, used to solve ordinary differential equations. Each variable in the equation corresponds to a shaft in the analyzer with the value of the variable being proportional to the number of rotations of the shaft. Mechanical connections among the shafts are used to enforce the desired mathematical relations among the variables. Calling the independent variable t and the dependent variables v_1, v_2, ..., v_M, Claude Shannon (1941) says that the system of equations is *solvable* by the analyzer just in case when its independent shaft t is turned, the dependent shafts v_1, v_2, ..., v_M turn in accord with the equations for any initial conditions. He says that a function f of $\mathbb{R}$ can be *generated* by the analyzer just in case it is a solution function $f(t) = v_i(t)$ for some i and some initial conditions.

If we abstract from the hardware, we are left with not much more than the idea of a system which is governed by laws L that are futuristically deterministic and time translation invariant (so that the development of the system does not depend on the instant at which the initial conditions are specified, as is discussed more fully in Ch. VII). If $(\mathring{v}_1, \mathring{v}_2, \ldots, \mathring{v}_M)$ is an allowed initial state at t_0, the uniquely determined solution functions $v_i(t)$, $t \in [t_0, +\infty)$, can be said to be analogue computable relative to the laws L. The justification for this terminology is that there is a computer — Nature herself — which computes the function as follows. We prepare the system in the state $(\mathring{v}_1, \mathring{v}_2, \ldots, \mathring{v}_M)$ at any instant t_0, wait $t - t_0$ seconds, and then read off the values displayed by the system.

Shannon (1941) and Pour-El (1974) consider laws in the form of first order ordinary differential equations representing the operation of a general purpose analogue computer which performs the operation of integrating a variable, multiplying a variable by a constant, and multi-plying and adding two variables. They show that hypertranscendental functions are not analogue computable (or generable in Shannon's terminology) by such a device. It follows that these devices are not capable of computing (or generating) some functions which are digitally computable by approximations, for some hypertranscendental func-tions, such as the reciprocal of the gamma function, are known to be Grzegorczyk computable. One would like to construct a more general analogue computer which would compute all Grzegorczyk computable functions, or else show why this is not possible.

Montague (1962) formalized the notion of a generalized computer which computes functions as follows. If x is an argument for which the function value is desired, the signal input variable w_1 is brought to its special starting value w^* and the argument input variable w_2 is brought to the value x. The computer then chugs away until the output signal variable w_3 flashes its special value w^{**} indicating that the simultaneous value y of the output variable w_4 is to be read as the value of the function for the argument x. This generalized computer may be run according to either deterministic or indeterministic laws. In either case the laws are assumed to be time translation invariant, and for simplicity it is assumed that if the signal output variable takes on its special value w^{**} then there is a first instant at which it takes it. The deterministic mode of operation requires of the laws L that for any histories, primed and unprimed, and any times t_0 and t_1, if $w_i'(t_0) = w_i(t_0)$, $i = 1, 2, 3, 4$, $w_1'(t_0) = w^*$ and if $t_1 \geqslant t_0$ is the first instant at which $w_3'(t_1) = w^{**}$, then $w_i'(t_1) = w_i(t_1)$, $i = 1, 2, 3, 4$. The value y is said to be computed for the argument x just in case there is an allowed history and times t_0 and t_1 such that $w_1(t_0) = w^*$, $w_2(t_0) = x$, $t_1 \geqslant t_0$ is the first instant at which $w_3(t_1) = w^{**}$, and $w_4(t_1) = y$. The function f is said to be computable by the computer just in case for every $x \in \text{dom}(f)$, $y = f(x)$ is the value computed for the argument x, and for every $x \notin \text{dom}(f)$ no value is computed for the argument x. In the indeterministic mode of operation it is required that for any allowed histories, primed and unprimed, and for any time t_0, if $w_1'(t_0) = w_1(t_0) = w^*$ and $w_2'(t_0) = w_2(t_0)$ and if $t_1 \geqslant t_0$ is the first instant such that $w_3(t_1) = w^{**}$, then there is also a first instant $t_2 \geqslant 0$ such that $w_3'(t_2) = w^{**}$ and further $w_4'(t_2) = w_4(t_1)$.

From the perspective of foundations it is useful to have a notion of computability which is general enough that the various notions of digital and analogue computability can all be obtained by specializing the variables and the laws L by which the generalized computer operates. That is what Montague's approach promises to provide, at least with a little fiddling which I will not attempt here. But the worry arises that the sense of computability involved is so general as to be useless; for if L is allowed to range over every kind of deterministic law then presumably few if any functions will fail to be generalized computable. The worry can be assuaged by emphasizing that the generalized concept of computability is relativized to laws L. The subject is given content by proving results about what functions are and are not generalized computable relative to what laws, especially the kinds of laws encountered in mathematical physics. As usual, I leave it to the reader to supply the labor.

8. OBJECTIONS; CHURCH'S THESIS REVISITED

The notions of generalized computability introduced in the preceding section are open to a series of objections which are worth reviewing because their resolution serves to clear away several potential misunderstandings. The consideration of the objections also serves as an opportunity to the reconsider the meaning of Church's thesis.

Objection 1. Your generalized 'computations' are presented in terms of a Big Computer in Sky. This mythical machine violates a basic part of any plausible definition of 'computer'. In his book *Analogue Computation*, Albert Jackson writes:

A computing device may be defined as a device that accepts quantitative information, arranges it and performs mathematical and logical operations on it, and makes a available resulting quantitative information as its outputs. (1960, p. 2)

But your Big Computer in the Sky does not make outputs available as information in the sense of symbols printed on paper and the like. *Response.* I view this aspect of computation as an engineering problem and not as part of the analysis of computability per se.

Objection 2. You are missing the point. The key idea is that a computer accepts and then outputs information. The exact definition of 'information' is not at issue here except in so far as it necessarily involves symbolic representation. Thus the maxim, 'No computation without symbolic representation', which is grossly flouted in your notions of generalized computability. *Response.* A definition of computable function might but need not make use of a coding of values by symbols. Whenever possible it seems best to characterize computability, digital or analogue, without reference to coding; for while no one can doubt that the standard Turing machine coding — a string of m consecutive '1's to code the positive integer m — is an effective coding, the general problem of distinguishing effective from non-effective codings is equivalent to the problem of deciding when a function is effectively computable, the very problem at issue. (It should be obvious, for example, that if the coding of integers is not effective, then any set of integers could be enumerated by a Turing machine.) In any case, when we move from an abstract generalized computer to a physical realization of it, there is representation; e.g., the variable v_3 (say) takes as values volts and r volts represents the real number r.

Objection 3. What we want of a representation is that it enables us to access the information, and the standard systems of symbolic representation, so conspicuously lacking in your characterization of generalized computability, are designed to guarantee such access. *Response.* Epistemic access is an important issue. And I would suggest that it is not merely the natural tendency to anthropomorphize when explaining the intuitive basis of computability but also the concern with epistemic access that explains the presence in Turing's original paper of such phrases as "scanning" a square, "immediate recognizability" of changes in squares, and "states of mind" of the computing agent. This concern also seems to be present in contemporary presentations of algorithms in terms of "a computing agent . . . which can react to the instructions and carry out the computations" (Rogers (1967), p. 2). But, to repeat, I do deny that computability is an epistemological concept. Turing computability can be presented in a purely abstract fashion, avoiding questions of representation and epistemic access; and just as the mathematical theory of Turing computability can be developed independently of these questions, so can the theory of generalized computability.

Objection 4. Granting for sake of argument that the generalized notions of computability do capture legitimate senses of effective, mechanical computability, these senses are so different from Turing's original sense that it is misleading to speak as if there were a unified concept of computability which covers all the bases. *Response.* It is and it isn't. It is useful to see Turing computability as a special case of a general notion of effective computability which covers digital and analogue computers and other deterministic devices as well. But Turing computability is such an important and distinct special case that it deserves special handling.

Start with the notion of a programmable or algorithmically computable function of the integers. Roughly, this is a function which is computable by means of a stepwise discrete procedure which can be carried out according to a finite list of instructions each of which . . . (For the ellipsis, fill in your favorite intuitive account.[6]) Church's thesis, or proposal as I would prefer to call it, says:

(CP1)   The class of programmable or algorithmically computable functions of the integers is to be identified with the Turing computable functions.

I have no doubts about the adequacy of (CP1)[7], especially as regards the originally intended application to Hilbert's decision problem. (Recall that Turing's original paper was entitled "On Computable Numbers, with an Application to the Entscheidungsproblem," and that Church's notion of effective calculability was introduced in the papers "An Unsolvable Problem of Elementary Number Theory" and "A Note on the Entscheidungsproblem.")

Church's initial proposal (CP1) could be extended to functions of the reals by

(CP2)   The class of programmable or algorithmically computable functions of the reals is to be identified with the Grzegorczyk computable functions.

However, (CP2) does not carry the conviction of (CP1) because Grzegorczyk's definition, though useful for proving results in analysis, is only one of various possible ways to generalize Turing computability to functions of the reals.

It is here that Turing's (monumental!) contribution ends. Turing is sometimes represented as having set and achieved the more ambitious goal of specifying the most general notion of what a 'machine' is and then using this notion to explicate the general notion of a mechanically computable function.[8] This corresponds to a third proposal.

(CP3)   The class of effectively, mechanically computable functions is to be identified with the class of programmable or algorithmically computable functions and, thus, with the Turing-Grzegorczyk computable functions.

Even leaving aside the qualms about (CP2) that infect (CP3), (CP3) is unacceptable, for it is simply wrong that a Turing machine is the most general type of machine that can perform what is recognizably an effective, mechanical computation of a function. What is true is that the theory of non-Turing computability remains to be developed. Whether the development along the lines suggested in Sec. 7 above will prove to be worth the effort remains to be seen.

9. CONCLUSION

Our starting question about the relation between determinism and mechanism can be given a partial answer. Determinism does not necessarily entail mechanism in the Turing-Grzegorczyk sense of effective computability, but various interesting partial entailments hold for many types of deterministic laws in the form of ordinary and partial differential equations. In the converse direction, effective, mechanical computability does not entail determinism, but any function which can be computed by an indeterministic Turing machine can also be computed by a deterministic Turing machine (though the computations of the latter may not be as efficient). In general, however, half of the question tends to collapse, for any deterministic and time translation invariant system can be regarded as an analogue computer which computes values of the solution functions. As for the other half of the question, an analogue computer need not operate deterministically, but whether the resort to indeterminism enlarges the class of analogue computable functions is a matter that has to be settled on a case by case basis.

NOTES

[1] Rogers (1967) and Tourlakis (1984) provide comprehensive treatments of this and other topics touched on in this section.

[2] For examples, see Rogers (1967).

[3] These machines were first discussed in Shepherdson and Sturgis (1963). A detailed development is given in Cutland (1980) whose presentation I follow here.

[4] For functions of the reals with domains confined to a closed and bounded interval with computable endpoints, clause (ii) simplifies to the requirement that there is a one-place Turing computable function $l: \mathbb{N} \to \mathbb{N}$ such that $|f(r) - f(r')| < 1/2^k$ whenever $|r - r'| < 1/l(k)$. Clause (i) corresponds to S. Mazur's (1963) definition of computable function of the reals. There are weaker versions of clause (ii) which give rise to notions

of computable function of the reals weaker than Grzegorczyk's. Grzegorczyk's definition has some nice consequences to which I will appeal to below. For example, if $\{f_k\}$ is a sequence of Grzegorczyk computable functions that converges uniformly and effectively to a function g, then g is Grzegorczyk computable; further, Riemann integration preserves Grzegorczyk computability.

[5] Or rather intuitions divide. The step function is clearly not computable in the sense of being 'programmable' since there is no recursive procedure to decide whether x equals x_0 (where x_0 is assumed to be a computable real).

[6] See, for example, Rogers (1967), Ch. 1.

[7] But for the record I note that others have expressed doubts in both directions, some questioning the idea that all recursive functions are effectively computable, and others questioning the converse; see Péter (1957), Bowie (1972), and Thomas (1973).

[8] I do not know whether this representation is historically accurate, though I suspect that it contains a kernel of truth. For example, because he thought of machine states on the analogy with states of mind, Turing (1937) ruled out machines in which the states can get arbitrarily close together, because otherwise the computing agent might confuse them.

SUGGESTED READINGS FOR CHAPTER VI

Alan Turing's (1937) original paper "Computable Numbers" still rewards the effort of reading. Cutland's (1980) *Computability*, Rogers' (1967) *Theory of Recursive Functions and Effective Computability*, and Tourlakis' (1984) *Computability* cover the standard topics in algorithmic computability for functions of the integers. For an introduction to computable analysis, see Aberth's (1980) book of that title. The series of papers by Pour-El and Richards cited in the text provide definitive results about the preservation of Turing-Grzegorczyk computability by linear partial differential equations. Pour-El's (1974) "Abstract Computability and Its Relation to the General Purpose Analogue Computer" presents a concept of analogue computability that covers the types of analogue computers actually in use.

CHAPTER VII

DETERMINISM AND TIME SYMMETRIES

> Now since infinite time must be assumed, no fresh
> possibility can exist and everything must have appeared
> already, and moreover an infinite number of times . . .
> (Friedrich Nietzsche, "Eternal Recurrence")

1. THE RECEIVED VIEW

Several of the classic treatments of determinism make various time symmetries part of the definition of determinism or else an immediate consequence of it. Recall that Russell's treatment appealed to a "uniformity of nature," meaning that "no scientific law involves time as an argument, unless, or course, it is given in integrated form, in which case *lapse* of time, though not absolute time, may appear in the formulae" (Russell (1953), p. 401). A similar sentiment is found in Herbert Feigl's discussion of the principle of causality:

The place and time at which events occur do not by themselves have any modifying effect on these events. Mathematically this may be expressed by saying that the space and time variables do not enter *explicitly* into the functions representing natural laws . . . 'Same causes, same effects' makes sense only if there is a neutral medium as space-time which is thus no more than a *principium individuationis*. Differences in effects must always be accounted for in terms of differences in *conditions*, not in terms of spatio-temporal location.[1]

Another kind of time symmetry involving eternal recurrence is found in John Stuart Mill's discussion of the "law of causation":

The state of the whole universe at any instant, we believe to be the consequent of its state at the previous instant . . . And if any particular state of the entire universe could ever recur a second time, all subsequent states would return too, and history would, like a circulating decimal of many figures, periodically repeat itself. (1904, pp. 400—401)

In an offshoot of the same tradition, Ernest Nagel also takes determinism to embody eternal recurrence in that if a system is brought back into the state it originally had at a given initial time t_0 and then is allowed to evolve of its own accord for an interval $t_1 - t_0$, it will exhibit at the end of that interval the same state as it originally had at t_1.[2]

It would be foolhardy to go against the combined authority of Mill, Russell, Feigl, and Nagel. Not being overly foolhardy, I will declare that our authorities are right — *partly* right. Before we can see in what part they are right, we need to define our terms a little more carefully. The next three sections discuss in turn three distinct but interconnected time symmetry properties.

2. TIME TRANSLATION INVARIANCE

To keep matters simple, I will restrict attention to classical physics. Only the technical details are changed when we make the transition to special relativistic physics. General relativistic physics, however, produces a major shift in the terms of the discussion; the reasons for the shift will be previewed at the end of this chapter and discussed more fully in Ch. X.

All of the possible worlds $\mathscr{W}_L$ allowed by the putative laws L are painted on the fixed canvas of Newtonian space-time $\langle \mathbb{R}^4, G_1, G_2, \ldots \rangle$, where $\mathbb{R}^4$ is the space-time manifold and the G_i are geometric object fields on $\mathbb{R}^4$ which characterize the geometric structure of space-time. Exactly what is to be included among the G's will have an important influence on the perception of time symmetries, but as a start let us assume the standard equipment listed in Ch. III. The definitions of the various time symmetries are made easier by using a time function $t: \mathbb{R}^4 \to \mathbb{R}$ whose level surfaces coincide with the planes of absolute simultaneity and whose differences coincide with temporal duration. It must be understood, however, that at this juncture t is an auxiliary device and that the origin $t = 0$ has no special geometrical or physical significance.

Time translation invariance demands that the physical possibilities are closed under the operation of time translation which shifts all of the physical contents of space-time forwards or backwards on time by a given amount.

> *Def.*   The laws L are *time translation invariant* just in case for any real Δ, if $W \in \mathscr{W}_L$, then also $W^\Delta \in \mathscr{W}_L$ where $W^\Delta(t) \equiv W(t + \Delta)$ for all times t.

(For the pedants: Take a world corresponding to $\langle\langle \mathbb{R}^4, G_1, G_2, \ldots \rangle,$ $\langle P_1, P_2, \ldots \rangle\rangle$ where the P_j are object fields describing the physical contents of the space-time $\langle \mathbb{R}^4, G_1, G_2, \ldots \rangle$ which is held fixed from world to world. We generate $W^\Delta = \langle\langle \mathbb{R}^4, G_1, G_2, \ldots \rangle, \langle P_1^\Delta, P_2^\Delta, \ldots \rangle\rangle$ by choosing a homomorphism h of the space-time (i.e., h is a diffeomorphism of $\mathbb{R}^4$ onto itself such that $h^*G_i = G_i$ for each i) such that $h \cdot t(x) = t(x) + \Delta, x \in \mathbb{R}^4$. Then $P_j = h^*P_j$.)

3. RECURRENCE: CONDITIONAL AND UNCONDITIONAL

Proofs of *un*conditional recurrence of a state of affairs can start either from determinism or its negation. Any finite state Laplacian deterministic system operating in discrete time must exhibit periodicities (see Ch. VIII). The famous Poincaré recurrence theorem for continuous state systems starts from Laplacian determinism and invokes finiteness and conservation of volume in phase space to show that for almost any given initial state and any $\varepsilon > 0$, the system will within a finite time (which depends on the choice of initial state and ε) return to within the specified ε of the starting state.[3] Alternatively, we can start from the assumption that the system evolves not deterministically but stochastically and then try to prove recurrence from conditions on the chances or probabilities that describe the non-deterministic tendencies of the system to pass from one state to another. We can achieve success if, for example, we assume the system to be a finite Markov process.[4]

The kind of recurrence of concern here is *con*ditional rather than unconditional. Mill and Nagel's claim is not that any state of the universe will recur but only that *if* it repeats it will do so eternally. This is made a bit more explicit in

Def.   The laws L are (conditionally) *periodic* just in case for any $W \in \mathscr{W}_L$ and any t_0, t_1, t_2, t_3, if $t_3 - t_2 = t_1 - t_0$ and if $W(t_2) = W(t_0)$ then $W(t_3) = W(t_1)$.

Of course, if an initial state can never repeat then our definition is vacuously satisfied and conditional periodicity loses its interest.

4. TIME REVERSAL INVARIANCE

The final time symmetry property to be considered here requires that the laws treat the past and future as symmetric mirror images of each other. More precisely, the physically possible worlds are closed under the operation of time reversal.

Def.   Suppose that the laws L are time translation invariant. They are then *time reversal invariant* just in case if $W \in \mathcal{W}_L$ then also $W^T \in \mathcal{W}_L$, where $W^T(t) = [W(-t)]^R$ with $[\,]^R$ being the state reversal operator.

What we really have is a definition schema rather than a definition. To turn the schema into a concrete definition requires that the action of $[\,]^R$ be specified. This cannot be done in advance of a specification of the ingredients of the instantaneous state description, and those ingredients will vary from case to case. But two formal properties of $[\,]^R$ are required to hold; viz., $[\,]^R$ is one-to-one on instantaneous states and it is involutory, i.e., $[[\cdot]^R]^R = [\cdot]$. For a scalar field Φ, it is usually assumed that $[\Phi(\mathbf{x}, t)]^R = \Phi(\mathbf{x}, t)$. For classical particle mechanics $[\,]^R$ acts by reversing the instantaneous velocities while leaving their positions fixed. In classical or relativistic electromagnetic theory the usual assumption is that $[\,]^R$ acts on electric and magnetic fields by $[\mathbf{E}(\mathbf{x}, t)]^R = \mathbf{E}(\mathbf{x}, t)$ and $[\mathbf{B}(\mathbf{x}, t)]^R = -\mathbf{B}(\mathbf{x}, t)$.[5] The required formal properties obviously obtain in all these examples.

This explanation of 'time reversal' makes it a misnomer since it is not time but rather motions and field strengths that are reversed. While this explanation has the virtue of removing the halo of mystery that hovers around the notion of time reversal, it suffers the defect of failing to provide an intrinsic prescription for obtaining W^T from W. Thus, for example, to obtain the reverse of particle motions we would have to choose a frame of reference and reverse three velocities relative to that frame. The alternative would be to obtain W^T from W by reversing both the time orientation of W and the four velocities of the particles. The details of how the time reversal operation is implemented do not affect the following discussion.

5. TIME REVERSAL INVARIANCE AND FUTURISTIC AND HISTORICAL DETERMINISM

From the intuitive meaning of time reversal invariance it should follow that for time reversal invariant laws L, futuristic and historical determinism stand or fall together. Let us verify this implication, showing for example that trying to assume the time reversal invariant laws L to be futuristically but not historically deterministic leads to contradiction. If historical determinism fails, then there are W_1, $W_2 \in \mathscr{W}_L$ and times t_1 and t_2, $t_1 < t_2$, such that $W_1(t_2) = W_2(t_2)$ but $W_1(t_1) \neq W_2(t_1)$. Since time translation invariance has been assumed, we can if necessary shift the time origin so that $t_1 < 0 < t_2$. By time reversal invariance W_1^T, $W_2^T \in \mathscr{W}_L$. Since $W_1^T(-t_2) = [W_1(+t_2)]^R$ and $W_2^T(-t_2) = [W_2(+t_2)]^R$ it follows that $W_1^T(-t_2) = W_2^T(-t_2)$. Similarly, from $W_1^T(-t_1) = [W_1(+t_1)]^R$ and $W_2^T(-t_1) = [W_2(+t_1)]^R$ and the properties of $[\ ]^R$ it follows that $W_1^T(-t_1) \neq W_2^T(-t_1)$. But this is a violation of futuristic determinism since $-t_1 > -t_2$ (see Fig. VII.1).

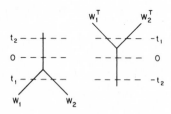

Fig. VII.1

While time reversal invariance is sufficient to guarantee that futuristic and historical determinism stand or fall together, it is not necessary. The classical heat equation studied in Ch. III provides a not entirely perfect example: it is not time reversal invariant, but with the temperatures at the ends of a finite rod stipulated, we have both futuristic and historical determinism. The modified heat equation with a non-linear term added (Ch. III.2) provides an example where futuristic determinism reigns but historical determinism fails. The reader should attempt to supply cleaner examples.

6. DETERMINISM PLUS TIME TRANSLATION INVARIANCE IMPLY PERIODICITY

We have seen that time translation invariance plus time reversal invariance implies that laws are deterministic in both directions of time if they are deterministic in either. Further, time translation invariance coupled with determinism guarantees conditional periodicity;[6] in fact, we are guaranteed the property that for any W_1, $W_2 \in \mathscr{W}_L$ and any times t_1, t_2 and any constant C, if $W_1(t_1) = W_2(t_2)$ then $W_1(t_1 + C) =$

$W_2(t_2 + C)$. Again the proof is by contradiction. Suppose that $W_1(t_1) = W_2(t_2)$ but $W_1(t_1 + C) \neq W_1(t_2 + C)$. By time translation invariance $W_1^\Delta \in \mathscr{W}_L$. Choose $\Delta = t_1 - t_2$. Then $W_1^\Delta(t_2) = W_1(t_1) = W_2(t_2)$. Similarly, $W_1^\Delta(t_2 + C) = W_1(t_1 + C) \neq W_2(t_2 + C)$, which is a violation of determinism.

7. MILL, RUSSELL, FEIGL, AND NAGEL: VINDICATED?

I said that our authorities were partly correct. It is now time to explain that half-hearted endorsement.

In standard Newtonian space-time time is absolute as regards simultaneity and duration, but there is no distinguished time origin. Time translation $t \to t + \Delta$ is thus a symmetry of the space-time. Applying our principle (Ch. III.2) that the symmetries of well-formulated laws must be as inclusive as the symmetries of the space-time on which they are based, we conclude that any laws based on standard Newtonian space-time must be time translation invariant.[7] This is the sense in which Russell and Feigl are correct. By the results of the previous section it follows further that any deterministic laws based on standard Newtonian space-time are conditionally periodic. This is the sense in which Mill and Nagel are correct.

8. MILL, RUSSELL, FEIGL, AND NAGEL: REFUTED?

If there is no higher law decreeing that classical laws must be based on standard Newtonian space-time, then the time translation symmetry of the space-time can be broken by introducing as part of the geometrical structure (the G's) a distinguished time function t^* whose zero identifies a special origin of time. By our principle (Ch. III.2) linking space-time symmetries and symmetries of laws, this introduction must go hand in hand with the breaking of time translation invariance for the laws of motion. Towards the latter, let us imagine that the usual Newtonian laws of gravitation are modified in favor of laws L^* which imply that the gravitational 'constant' K is not constant but a smoothly varying function of time. If the new laws L^* specify the time behavior $K = K(t^*)$, then they should count as being every bit as deterministic as the old laws. But if the laws L^* allow different Newtonian position-velocity states to coexist with the same K value then both time translation invariance and periodicity seem to fail since the time of obtaining of the 'initial' positions and velocities of the gravitating particles can make a difference in the subsequent motions.

9. RUSSELL AND FEIGL: DEFENDED

In the face of such examples we may be met with a reiteration of Feigl's claim that differences in effects must be accounted for in terms of differences in conditions, not in terms of differences in spatio-temporal locations. In the case in point, Feigl would insist that the differences in the temporal distances from the origin $t^* = 0$ don't in themselves make for differences in the particle motions; it is rather differences in the values of the gravitational 'constant' K that account for differences in the motions. In other words, the value of K is part of the conditions, and since this value varies it must be included in the state description. And under this more comprehensive state description, time translation invariance and periodicity are restored.

10. MILL, RUSSELL, FEIGL, AND NAGEL: MODIFIED AND QUALIFIED

Though there is something appealing about the move just made to defend Russell and Feigl, there is also something more than a little disturbing about the implications which emerge when the defense is pushed to its logical limits. If differences in temporal evolution are always chalked up to differences in 'conditions', then time translation invariance is an *a priori* property of natural laws. This conclusion does not sit well with an empiricist analysis which would explain symmetry principles as deep but contingent laws. Even worse, if there is no limit to what we are willing to pack into the category of the physical contents of space-time (the P's as opposed to the G's) in order to satisfy Feigl's demand, then periodicity may be trivialized. For instance, if the function t^* is said to belong to P's instead of G's, on the grounds that it is a scalar physical field on a par with the other physical fields, then no state can repeat itself and periodicity is automatically and vacuously fulfilled.

It is conceivable that careful observations on quasi-isolated systems could rule out any plausible source equation for K and confirm instead that the temporal evolution of K is independent of the amount and distribution of gravitating matter and of any other known 'physical' source. In that case denying to K the status of a Feigl 'condition' is not only not an *ad hoc* move but is supported by the principle that the cut between what goes into the geometrical structure of space-time itself and what goes into the physical contents of space-time corresponds to the cut between absolute and dynamical objects, the former being objects than remain fixed in every physically possible world and the latter being objects that vary from physically possible world to world.[8] Recall Newton's dictum that "Absolute space, in its own nature, without relation to anything external, remains always similar and immovable". Substitute 'space-time' for 'space' and we have exactly the distinction between the G's and the P's in our orthodox classical world models. Moreover, the corresponding distinction remains valid for special relativistic worlds as well.

I conclude that, contrary to Russell and Feigl, there are conceivable circumstances in which the most natural and plausible description would violate time translation invariance; and that, contrary to Mill and Nagel, the same circumstances would also violate periodicity without necessarily violating determinism. Such circumstances are farfetched in that they are unlike anything found in the history of physics, and had they arisen they would have caused an upheaval in the classical and special relativistic conceptions of the nature of space-time. But it is Nature and not philosophers who must decide whether or not to throw such circumstances at us. There is no *a priori* prohibition against them to be derived from conceptual truths about space, time, and determinism.

11. PREVIEW: GENERAL RELATIVITY AND TIME SYMMETRIES

In general relativistic physics both of the distinctions geometrical space-time structures vs. physical contents of space-time and absolute vs. dynamical objects tend to collapse. With this collapse time translation invariance as it was formulated above for classical and special relativistic theories becomes moot.[9] We can still talk about recurrence; but unconditional Poincaré type recurrence may fail because time itself runs out before the initial state has a chance to recur, and for the same reason conditional recurrence can trivialize. Tipler (1980) shows that in a spatially finite and deterministic general relativistic universe,[10] the initial state cannot recur if the universe begins to undergo gravitation collapse. This impossibility result does not hold for spatially infinite universes.

NOTES

[1] Feigl (1953), p. 412. See also J. C. Maxwell (1920), whom Feigl cites as his authority for the principle of the 'homogeneity of time'.

[2] "Suppose then at time t_0, [the system] S is in a state describable as P_0, Q_0, R_0, etc., that the state of S changes with time, and that at time t_1 S is in a state describable as P_1, Q_1, R_1, etc. Next imagine that S is brought back into the state it originally possessed at time t_0, that it is then permitted to change of its own accord, and that after an interval $t_1 - t_0$ it once more exhibits the state describable as P_1, Q_1, R_1, that is, its state is once more what it was at time t_1. A system which always behaved in this manner would be one in which its state at one time uniquely determined its state at any other time"; E. Nagel (1953, pp. 421—422). Nagel is attempting to use periodicity to define determinism; this obviously won't do if states cannot repeat. What is more plausible, and what Nagel means to imply, is that determinism implies conditional periodicity. See also E. Nagel (1961, pp. 279—280).

[3] See Ch. IX below for a more detailed discussion of this and related matters.

[4] See Feller (1968), Ch. XVI. Nietzsche would have seen this as a proof of his ideas about eternal recurrence, as noted in Tipler (1980).

[5] In ordinary quantum mechanics, $[\psi(x, t)]^R = \psi^*(x, t)$, where ψ is the state function of a spinless particle and '*' denotes complex conjugation; see Ch. XI below. There are many interesting problems about the physical motivation for the choice of $[\ ]^R$ in particular cases, but the reader will have to explore these issues on his own.

[6] Montague (1974) showed that conditional periodicity does not follow from determinism *per se*. What is now at issue, however, is whether the implication holds if 'plausible' assumptions about the nature of space and time are added. This issue is addressed in the next four sections.

[7] For classical mechanical systems whose dynamics can be cast into Hamiltonian form (see Ch. IX below) there is a tight connection between spatio-temporal symmetries and conservation laws. In the case in point, time translation invariance holds if and only if there is conservation of energy. Thus, for Hamiltonian systems, to the extent that time translation invariance is secure, so is conservation of energy, and vice versa. In non-Hamiltonian systems there is not always such a neat connection between symmetry and conservation.

[8] See J. L. Anderson (1967) and M. Friedman (1983) for attempts to make this distinction precise.

[9] See R. Jones (1981) for an interesting discussion of these issues.

[10] In the language of Ch. X below, the space-time possesses a compact Cauchy surface.

SUGGESTED READINGS FOR CHAPTER VII

Montague's (1974) paper "Deterministic Theories" injected some much needed precision into the discussion of the connection between determinism and the symmetries of time translation invariance and (conditional) periodicity. But I have not been able to locate a good general discussion of the relation of determinism to time symmetries which shows how the relationship affects and is affected by space-time structure. There are any number of technical treatises on symmetries and invariances in various branches of physics; see, for example, Elliott and Dawber (1979) *Symmetry in Physics*.

CHAPTER VIII

DETERMINISM, RANDOMNESS, AND CHAOS

> Too keen an eye for pattern will find it anywhere.
> (T. L. Fine, *Theories of Probability*)

'Deterministic' is variously contrasted with 'random', 'disordered', 'stochastic', 'chancy', 'capricious', and 'haphazard'. If the contrasts were true and sharp they might be used to enhance our understanding of determinism. But each of the alleged contrast terms is as much in need of analysis as is determinism. Certainly ordinary usage is confusing in its incestuous interweaving of the terms. Thus, the *Oxford English Dictionary* defines 'random' as "at haphazard, without aim, purpose or fixed principle"; 'chaotic' as "utterly confused or disordered"; and 'haphazard' as "by mere chance, without design; at random".

To bring some order to this confused situation I will propose a not altogether arbitrary grouping of these terms. Within each group I will distinguish a process sense from a performance sense, and I will attend to distinctions among levels of description. Thus, for present purposes I will group together chaotic, capricious, and haphazard. A chaotic, capricious, or haphazard process I declare to be one which operates "without fixed principle," without the guidance of natural laws whether deterministic or not. The output of a process is chaotic in the performance sense if it is disordered or lacking in pattern. Process chaos, I hold, is a coherent notion while utter chaos in the performance sense is not. And attending to the levels distinction, we must be prepared to find that chaos or caprice at one level of description gives way to order and design at another level, or vice versa. Against the *OED*, I group random with stochastic or chancy, taking a random process to be one which does not operate wholly capriciously or haphazardly but in accord with stochastic or probabilistic laws.[1] Whether there are in nature ultimately random processes, processes that remain random no matter how deep the level of description is pushed, is a controversial question that must be broached in discussing quantum physics (see Ch. XI). But in the present chapter the focus will be largely on performance randomness at a level of description where the process is to all appearances stochastic.

137

Randomness in performance is not an absolute concept, for which features we look for in a random dance will depend upon what we know or believe about the animating stochastic process. Nor is performance randomness inevitable in the outputs of a random process, though it is 'likely'. And randomness in performance does not always speak truly of genesis randomness. These are some of the points I will try to clarify in the coming sections.

Much of the early discussion of randomness derived from a wrong-headed view of probability — the so-called frequency theory. Rather than trace the tangled history of this subject, let us begin more constructively; contact with the historical tradition will be established once we have laid the basic framework for an understanding of the issues.

1. DEFINING RANDOMNESS

Consider a binary state process operating in discrete time. Letting '0' and '1' denote the two possible states, a history of the process is a map $s: \mathbb{Z} \rightarrow \{0, 1\}$, $s(n)$ being the state at time t_n. Corresponding to a history is a doubly infinite binary sequence.

$$\ldots s(-2), s(-1), s(0), s(+1), s(+2), \ldots$$

The collection of all such sequences is denoted by X^∞ and the collection of all initial finite segments starting from some place m and moving into the future is denoted by X_m^*.

A *stationary stochastic model* for our binary process assigns to finite segments of member sequences of X^∞ probabilities that are independent of their temporal location. To make this precise, let X^n denote the collection of the 2^n binary sequences of length $n > 0$. For each n, the model assigns a probability function Pr_n to X^n, with the probabilities at different levels meshing according to

$$\mathrm{Pr}_n(x) = \mathrm{Pr}_{n+1}(x0) + \mathrm{Pr}_{n+1}(x1), \quad x \in X^n$$

Then for any m, the stationary probability for the finite stretch of history s from time t_m to t_{m+n} is

$$\mathrm{Pr}_n((s(m), s(m + 1), \ldots, s(m + n)).$$

The subscripts on Pr will henceforth be dropped.

The most familiar example of a stationary stochastic model is provided by a Bernoulli process where the successive states are probabilistically independent and the probabilities for 0 and 1 have (constant) values p and $1 - p$ respectively. Then the probability for any finite segment of length n containing n_1 0's and n_2 1's is $p^{n_1}(1 - p)^{n_2} = p^{n_1}(1 - p)^{n - n_1}$. In the special case where $p = 1 - p = \frac{1}{2}$ ('fair coin') every segment of length n has probability $1/2^n$.

Starting from our Pr assignments we can construct a measure μ on subsets of X^∞. The first step is to define μ on the cylinder sets $X_{mn} \subset X^\infty$ consisting of all elements of X^∞ agreeing in places m through n $(m \leqslant n)$:

$$\mu(X_{mn}) = \Pr((s(m), s(m + 1), \ldots, s(n))$$

where s is any one of the agreeing histories. The measure is then extended in the obvious way to the algebra of subsets of X^∞ generated by the cylinder subsets. And assuming countable additivity,[2] there is a unique extension to the full sigma-algebra of subsets of X^∞.

The reader who has endured these definitions may have begun to wonder where the randomness is. There is more — and less — here that meets the eye! Consider the two infinite sequences:

(a) $\ldots 0, 1, 0, 1, 0, 1, 0, 1, 0, 1, \ldots$

(b) $\ldots 0, 1, 1, 0, 1, 0, 0, 1, 1, 0, \ldots$

Intuitively, (b) is random while (a) is not. But with $p = \frac{1}{2}$ in a Bernoulli process, the pictured finite segments of (a) and (b) both have the probability $1/2^{10}$, and both of the singleton sets consisting of (a) and of (b) alone have μ measure 0. So it would seem that either from the finite or the infinite perspective, both cases are equally likely or equally unlikely, though they differ in randomness.

Nevertheless, there is a sense in which infinite random sequences are the likely outcomes of a random ($=$ stochastic) process while the non-random sequences are the unlikely ones; namely, if $\mathscr{R}, \mathscr{N} \subset X^\infty$ are respectively the collections of all random and all non-random sequences, then

(C) $\mu(\mathscr{R}) = 1, \quad \mu(\mathscr{N}) = 0.$

The problem of defining randomness now boils down to choosing the appropriate sets of measure 1 (or of measure 0).

Start at some time, say t_0, and move towards the future. We will then be concerned for the moment with the singly infinite sequences X_0^∞, obtained by chopping off the past histories, and with the initial finite segments X_0^*. If the process is Bernoulli with probability p for '0', we would expect that as we scan longer and longer stretches of some $x \in X_0^\infty$ the relative frequency of 0's will tend towards p. The strong form of the law of large numbers shows that our expectations are correct in the sense of μ-measure: the μ-measure of the set of X_0^∞ sequences in which the relative frequencies converges to p is 1. Having proved this we will certainly want to have $\mathscr{R}$ be a subset of this set of measure 1. And $\mathscr{R}$ will be a proper subset at that. For we can divide the sequences X_0^∞ into segments of length, say, 4 and track the relative frequency of, say, $(1, 0, 1, 1)$. Our expectations are that the frequency will tend towards $p(1 - p)^3$. Again, mathematics confirms our expectations in that the collection of all sequences in which the right convergence takes place is also of μ-measure 1. Now we will want $\mathscr{R}$ to be in the intersection of these two nice measure 1 sets. We can go on, and on, and on in this way, but we must draw the line somewhere, else we will end up taking $\mathscr{R}$ to be in the intersection of all subsets of X_0^∞ of μ-measure 1, leading to $\mathscr{R} = \emptyset$ and violating (C).

There are two rather different attitudes towards this problem. The first is to be liberal about what counts as a property of randomness and, consequently, stingy about what counts as a random sequence. Thus, Wald (1938) and Martin-Löf (1970) have proposed to count as a property of randomness any property which has μ-measure one and which is expressible in a certain way. Wald's version requires expressibility in a standard logic such as *Principia Mathematica*. Martin-Löf's version requires that the properties be hyperarithmetical (roughly, these are the properties expressible in an infinitary propositional calculus). He shows that the intersection of all hyperarithmetical sets of μ-measure 1 is also of measure 1 so that (C) is satisfied for the Martin-Löf random sequences $\mathscr{R}_{ML}$.

The opposed strategy is to be more stingy about what counts as a property of randomness and, thus, more liberal about what counts as a random sequence. The classic example of this approach arises from von Mises' (1957, 1964) theory of probability as the study of 'Kollektives'. Binary Kollektives are essentially subsets of X_0^∞ in which the limiting relative frequencies of 0's and 1's converge. The limiting frequencies are required to be invariant under 'place selection rules'. Thinking of the original application of probability theory as a guide to betting behavior where a bet on the outcome of the $n + 1^{st}$ trial is made after witnessing the first n trials, we can take a place selection rule to be a function f from X_0^* to $\{0, 1\}$. Applying the rule f to a member sequence x of the Kollektives, the $n + 1^{st}$ item is selected or rejected according as $f(x(n)) = 1$ or 0 where $x(n)$ is the initial segment of length n of x. Carrying the gambling analogy further, a successful gambling system requires that an effective procedure for selecting the items on which bets are to be laid. This led Church (1940) to suggest that the selection rules should be Turing computable functions. Thus, a sequence $x \in X_0^\infty$ is Church random just in case the limiting relative frequency of 0's converges to some value, say $\frac{1}{2}$, and the same limiting value exists in every infinite subsequence of x produced by the application of a recursive selection rule. Since there are only a countable number of selection rules in the von Mises-Church sense the existence of Church random sequences is assured, and the application of the μ-measure for a Bernoulli process with $p = \frac{1}{2}$ verifies that the collection $\mathscr{R}_C$ of all Church random sequences fulfills (C).

It is known, however, that $\mathscr{R}_C$ is too broad. It follows from the results of Ville (1939) that $\mathscr{R}_C$ contains sequences in which the relative frequency of 0's in every initial finite segment is greater than or equal to $\frac{1}{2}$. The observation of such a frequency behavior would surely lead us to question the Bernoulli model with $p = \frac{1}{2}$.

More adequate versions of the definition of infinite random sequences as those which pass effective statistical tests[3] have been offered by Martin-Löf (1966), Schnorr (1971, 1971a), and T. Fine (1973). The different versions yield somewhat different $\mathscr{R}_E s$, but in general we have $\mathscr{R}_{ML} \subset \mathscr{R}_E \subset \mathscr{R}_C$, where $\subset$ is proper inclusion.

2. RANDOMNESS, DISORDER, AND COMPUTATIONAL COMPLEXITY

The discussion so far has been defective in two ways: it has been silent on what randomness for finite sequences means and it has not related randomness to the intuitive notion of disorder. The concept of computational complexity promises to overcome both defects.

If A is a mechanical algorithm with $k + 1$ inputs and $x \in X_0^*$, then define the conditional complexity $K_A(x/i_1, i_2, \ldots, i_k)$ of x on the information $i_1, i_2, \ldots, i_k$ to be the length of the shortest 'program' P which joined to the information allows x to be computed, i.e., $A(P, i_1, i_2, \ldots, i_k) = x$; if there is no such P, then the complexity is $+\infty$. Conditional complexity depends on the choice of algorithm, but much of the arbitrariness can be removed by using Kolmogorov's result that there is an A^* such that

$$K_{A^*}(x/i_1, i_2, \ldots, i_k) \leqslant K_A(x/i_1, i_2, \ldots, i_k) + c$$

where the constant c may depend on A and A^* but is independent of x and the i's. It follows that

$$K_{A^*}(x/i_1, i_2, \ldots, i_k) \leqslant \log_2 N(i_1, i_2, \ldots, i_k) + c$$

where $N(i_1, i_2, \ldots, i_k)$ is the number of sequences for which the information holds. In the special case where the information contains only the length $l(x)$ of x, $N = 2^{l(x)}$ and $K_{A^*}(x/l(x)) \leqslant l(x) + c$. From here on the subscript on the complexity measure will be dropped.

Thus, for a finite sequence x we can say that x is random just in case it has high complexity in that

(C$_1$) $K(x/l(x)) \geqslant l(x) - d$, d a non-negative constant.

The choice of d is somewhat arbitrary, and thus so is the collection of finite random sequences.

Note that although the complexity notion of randomness is based on effective (digital) computability, there is no general effective procedure for computing the complexity of finite sequences and, thus, for deciding whether an arbitrary finite sequence is random. This fact is a corollary of the lemma that deciding whether any given Turing machine halts for any specified input is a recursively unsolvable problem.

Extending the computational complexity idea to infinite sequences can be done in a number of ways, only one of which will be considered here. For $x \in X_0^\infty$ again let $x(n)$ be the initial finite segment of length n. We can call x complexity random just in case

(C$_2$) $K(x(n)/n) \geqslant n - d$ infinitely often as $n \to \infty$.[4]

For the μ-measure derived from the Bernoulli model with $p = \frac{1}{2}$ the collection of all complexity random sequences $\mathscr{R}_{com}$ has measure 1 in accord with (C).

For infinite sequences the notion of randomness as complexity does not coincide with the notion of randomness as passing effective statistical tests; in fact, $\mathscr{R}_{com}$ is a proper subset of $\mathscr{R}_E$ on various definitions of effective test. This has led Schnorr (1971, 1971a) to reject the complexity approach on the grounds that it imposes properties of randomness that have no physical meaning in that the failure of such a property could never be detected by effective means.

There is another tension between randomness and complexity which needs to be addressed. The tension can be exposed by flipping a bent coin.

3. BIASED COINS

The computational complexity definition of randomness appears to fit within the ambit of the approach adopted in Sec. 1, but that appearance is an artifact of the choice of the value $p = \frac{1}{2}$. The root notion of randomness in Sec. 1 is that a random output process is one which reflects the probabilities of the underlying random ($=$ stochastic) process. But the complexity approach is absolute in that it makes no allowance for different p values. Consider a coin tossing experiment with a bent coin strongly biased in favor of heads (0) as against tails (1), with $p = 3/4$. Then for large n we would expect that about 3/4 of the flips to land heads, and in such a sequence the maximum (length relativized) complexity is about $(4/5)n$. Thus, the sequence will be rejected by criterion (C_1) even though the $(3/4)n$ heads are sprinkled in as random an arrangement as could be hoped. Conversely, finite sequences satisfying (C_1) will tend to be rejected by statistical tests postulating the p-value of 3/4.

It might be replied that flipping a bent coin is not a random experiment. But such an attitude seems to me to reveal an unjustified bias in favor of $p = \frac{1}{2}$, and it severely limits the application of the concept of randomness. As a kind of compromise, the complexity measure could be relativized not only to the length $l(x)$ of a sequence $x \in X_0^\infty$ but also to the weight $w(x)$, defined as the number of 1s in the sequence. Since

$N(l(x), w(x)) = \begin{pmatrix} l(x) \\ w(x) \end{pmatrix}$ the new weight relativized complexity criterion

for finite sequences is

$$(C_1') \qquad K(x/l(x), w(x)) \geq \log_2 \begin{pmatrix} l(x) \\ w(x) \end{pmatrix} - d$$

and the weight relativized counterpart of (C_2) for infinite sequences is

$$(C_2') \qquad K(x(n)/n, w(x(n))) \geq \log_2 \begin{pmatrix} l(x) \\ w(x) \end{pmatrix} - d$$

infinitely often as $n \to \infty$.

Liberalizing the complexity criteria in this way puts us on a slippery slope. If we relativize to weight, then why not to other features of the sequence as well? But the more we relativize the more sequences that are countered as random until finally no sequence is excluded on the complexity approach. However the line is drawn the fact remains that the contemplated relativizations do not properly take into account the underlying p-value. (C_1') and (C_2') will count as random some finite and infinite sequences which display a frequency of heads of 3/4, as would be expected when $p = 3/4$. But they will also count as random some sequences in which the frequency of heads is markedly different from 3/4, a result which is contrary to the root notion of Sec. 1 if indeed $p = 3/4$.

A seeming virtue of the complexity approach is that it gives a criterion of randomness for finite sequences that corresponds well to the intuitive notion of disorder. But the criterion is wrong from the point of view that a random performance is one which reflects the probabilities of the underlying stochastic process. This point of view, however, does not have a satisfying link to the notion of disorder for finite sequences. We could say that a finite output sequence from a Bernoulli process is random with respect to a p-value just in case the sequence passes statistical tests on the hypothesis that p has the stated value; that is, we do statistical hypothesis testing but reject sequences rather than the hypothesis.[5] But for $p = \frac{1}{2}$ all finite sequences are equally likely, so all sequences or none will pass regardless of how 'ordered' or 'disordered' the sprinkling of heads and tails. And for any p-value, the independence of trials means that all sequences with the same number of heads and tails are equally likely, so all or none of these sequences will pass, again regardless of how 'ordered' or 'disordered' the sequence.[6]

We have to face the fact that there are two concepts of performance randomness, or as I would prefer to say, the fact that there is a concept of randomness and a separable concept of disorder. The concept of disorder is an intrinsic notion; it takes the sequence at face value, caring nothing for genesis, and asks whether the sequence lacks pattern. The various complexity measures, relativized to length, weight, etc., provide different explications of this notion of disorder or pattern-freeness. By contrast, the concept of randomness is concerned with genesis; it does not take the sequence at face value but asks whether the sequence mirrors the probabilities of the process of which it is a product. There is a connection between this concept of randomness and the concept of disorder, but it is not a tight one. The various explications of randomness for infinite sequences guarantee the absence of pattern, but not in as strong a sense as the computational complexity approach; and for finite sequences the connection is looser still, with no guarantee that a sequence that passes statistical tests for a postulated p-value will be disordered in the complexity sense.

4. UTTER CHAOS

Chaos in the genesis sense — the absence of any lawlike guiding principle — is, I would argue, at least a conceptual possibility (see Ch. V, Sec. 6). But performance chaos — the absence of any pattern or order — may not be a coherent notion. Taking high complexity to mean low order and lack of pattern, we might try to define $x \in X_0^\infty$ to be *utterly chaotic* just in case

$$(C_3) \qquad K(x(n)/n) \geq n - e \qquad \text{for all } n \text{ (e non-negative constant)}$$

It has been shown by Martin-Löf that there is no such sequence! It follows, of course, that there is no sequence for which the inequality in (C_3) holds for all but a finite number of initial segments. The next most chaotic thing would be for the inequality to hold infinitely often, i.e., the sequence is random in the complexity sense; but this minimal form of chaos implies that the relative frequency of 0's converges to a limit. Out of chaos order arises!

A similar result holds for the weight relativized complexity measure on finite sequences. If (C_1') holds then the sequence $x = (x_1, x_2, \ldots, x_{l(x)})$ seems to converge in that

$$\max_{m \leq j \leq l(x)} \left| \frac{1}{j} \sum_i x_i - \frac{w(x)}{l(x)} \right| < \varepsilon,$$

where $\varepsilon = \left(\dfrac{2^{d+c}}{m} \right)^{1/4}$ (see T. Fine (1973)).

Rather than give up on the notion of utter performance chaos one may choose to abandon the computational complexity approach to it. The challenge is then to find an alternative approach and to demonstrate that it makes coherent sense of utter and complete performance chaos. I predict that the challenge cannot be met.

5. DETERMINISM AND PERFORMANCE RANDOMNESS

Genesis randomness is no sure guarantee of performance randomness, though in the μ-measure sense it leads to a strong expectation of a random dance. Conversely, randomness or disorder in a performance of finite length is no guarantee of genesis randomness, as is illustrated by 'random number' generators in digital computers.

However, there are conflicts between determinism and randomness in infinite sequences for discrete state devices. Consider a time translation invariant finite state device operating in discrete time. If the state at any instant is uniquely determined by the states at the preceding K instants, K finite, then any infinite sequence of states that constitutes an allowed history will exhibit periodicities which, on any reasonable definition of randomness, will brand the sequence as non-random. For a binary state device, for example, the maximum period is of length 2^K.

To bypass this difficulty, let us allow that the future state may depend on the entire infinite past history. Call a collection $D \subset X^\infty$ weakly futuristically (respectively, historically) deterministic just in case for any $x, x' \in D$ and any $m \in \mathbb{Z}$, if $x_i' = x_i$ for all $i \leq m$ (respectively, for all $i \geq m$), then $x' = x$. For any chosen criterion of performance randomness in infinite sequences, there will be weakly deterministic collections D some or all of whose member sequences are random. (For a doubly infinite sequence x we take randomness to mean that each of the singly infinite sequences $x_m, x_{m+1}, \ldots$, starting from any $m \in \mathbb{Z}$ is random.) But there is a more subtle conflict between weak determinism and performance randomness. A binary state deterministic device is severely limited in its ability to mimic the outputs of a random (= stochastic) process in that the possible output sequences D can cover only an insignificant fraction of the possible random sequences $\mathscr{R}$ which could result from, say, a Bernoulli process; for $\mu(D) = 0$ for *any* weakly deterministic D and *any* Bernoulli based μ-measure, with $0 < p < 1$.[7] To prove this we will assume without any real loss of generality that $p = \frac{1}{2}$. Divide the collection D into two parts D^- and D^+ according as there is a 0 or a 1 in the 0th place of the sequence. Let D^{-*} be the result of replacing 0 by 1 at the 0th place in each of the member sequences of D^-. Then $D^{-*} \cap D^+ = \emptyset$ since from the definition of weak determinism two member sequences of D cannot differ in only one place. Further, from the construction of the Bernoulli measure, $\mu(D^{-*}) = \mu(D^-)$ and $\mu(D^{-*} \cup D^+) \leqslant \frac{1}{2}$. Thus, $\mu(D^{-*} \cup D^+) = \mu(D^- \cup D^+) = \mu(D) \leqslant \frac{1}{2}$. Now iterate the procedure, starting from the deterministic collection $D_1 = D^{-*} \cup D^+$. Divide D_1 into D_1^- and D_1^+ according as there is a 0 or 1 in the first place, and let D_1^{-*} be the result of replacing 0 by 1 at the 1st place in each of the member sequences of D_1. Then $\mu(D_1^{-*} \cup D_1^+) = \mu(D_1^- \cup D_1^+) = \mu(D) \leqslant 1/4$. And in general $\mu(D_n) = \mu(D) \leqslant 1/2^n$, with the result that $\mu(D) = 0$. A similar conflict arises when we generalize from binary $0-1$ states to any finite number of states.

If we go to continuous state systems operating in continuous time and permit a shift in levels of description, the conflict between determinism and performance randomness can disappear. Indeed, such systems can develop in a strictly Laplacian deterministic manner, and yet on a coarse-grained level of description almost every coarse-grained sequence of states can be random and every such random sequence will be generated by an underlying deterministic sequence. These matters will be taken up in the following chapter.

6. PHYSICAL PROBABILITIES: FREQUENCIES OR PROPENSITIES?

While there is no disagreement about how to manipulate the formalism of stochastic models, the interpretation of the formalism has been and continues to be one of the most contentious areas in philosophy of

science. I will simply ignore the contentions of those probability
theorists who assert that all probability statements ultimately refer to
subjective or personal degrees of belief. The theories of physics purport
to give us truths about probabilities that are objective and observer
independent, and it is the nature of these probabilities we seek to
understand. The traditional battlelines of philosophical interpretation
pit the propensity theorist against the frequency theorist.[8]

To call probabilities propensities is only to attach a label that does
not explain anything. And worse, attaching that label raises the fears
that non-Humean powers are being attributed to Nature. Recall from
Ch. V the basic Empiricist demand on laws:

(E_1) For any worlds W_1 and W_2, if W_1 and W_2 agree on all
            occurrent facts, then they agree on laws.

If we are confident that non-probabilistic laws are Empiricist in this
sense, then we will also be confident of

(D) For any worlds W_1 and W_2, if W_1 and W_2 agree on all
            occurrent facts, then they agree on non-probabilistic disposi-
            tions, such as solubility.

For, presumably, occurrent facts about the microstructure of salts and
crystals together with the Empiricist laws will determine everything that
is true about the solubility of these materials. An Empiricist will also
insist that (D) continues to hold when extended beyond garden variety
dispositions such as solubility to the more exotic species of probabilistic
propensities:

(D') For any worlds W_1 and W_2, if W_1 and W_2 agree on all
            occurrent facts, then they agree on all physical probabilities.

But it is far from clear why (D') should hold on the 'single case'
propensity interpretation of probability; indeed, it is clear that (D') will
fail if the only relevant occurrent facts are those about the outcomes of
chance experiments. If two Bernoulli processes disagree on p-values, it
is unlikely that their output sequences will agree — unlikely in any
particular case but perfectly possible, and inevitable within the full
range of possibilities. Of course, different p-values may be grounded on
other kinds of occurrent facts in such away that (D') is satisfied. But it
is up to the propensity theorist to show how this grounding is achieved.
Not only do we get no help from the propensity theorist, but the avowal

that propensities involve a "new physical hypothesis (or perhaps a new metaphysical hypothesis)" (Popper, 1959) suggests that no such grounding is possible.

Frequency theorists disavow any new physical or metaphysical categories and propose to analyze physical probabilities in such a way that (D') is satisfied. The goal is worthy but the analysis is both unworkable and wrongheaded. Defining probability as the actual limiting frequency fulfills (D') but it makes the probability concept virtually inapplicable since actually infinite sequences of trials rarely if ever exist. Thus, a resort to hypothetical sequences of trials is inevitable. A resort to hypothetical outcomes is already familiar from the standard analysis of non-probabilistic dispositions such as solubility. That object o is soluble is taken to mean that o would dissolve if it were put into water; or in possible world talk, "o is soluble" is true (in the actual world $W_@$) just in case in every world W which is physically possible (relative to $W_@$) and in which o is immersed in water, o dissolves. If 'W is physically relative to $W_@$' means that W satisfies the laws of $W_@$ and if laws fulfill (E_1), then this analysis of solubility guarantees that (D) is fulfilled. Hypothetical frequency theorists, hoping to satisfy (D'), ape this approach, using hypothetical occurrent frequencies in place of hypothetical occurrent dissolvings. Call a world W an infinite future extension (ife) of $W_@$ just in case W agrees with the actual (finite) outcomes of the chance experiment in question and also W extends the repetitions of the experiment *ad infinitum*. Then the hypothetical frequency analysis would take "The probability for the outcome '0' is p" to be true (in $W_@$) just in case in every (or almost every) ife W which is physically possible (relative to $W_@$), the limit of the relative frequency of 0s is p.[9] The strong ('every') reading is unacceptable, for any reasonable interpretation of physical probability has to allow for the physical possibility that the frequency fails to converge or else converges to a value different from the true p-value. The weak ('almost every') reading escapes this difficulty, but it provides no definite truth conditions until the crucial 'almost every' clause is cashed in. Fetzer and Nute (1979, 1980) suggest a way to cash in frequency terms. Let W_1, $W_2, \ldots$ be an infinite sequence of ife worlds each of which is physically possible relative to $W_@$. The 'almost every' is taken to mean that $\lim_{n \to \infty} \#(W_n)/n = 1$, where $\#(W_n)$ is the number of worlds among the first n in which the limiting frequency of 0s is p. But if there are an infinite number of physically possible worlds in which the frequency converges to p there are surely also an infinite number in which the frequency doesn't converge or else converges to a value $p' \neq p$. Thus, satisfaction of the truth conditions for a given p-value will depend on the choice of the ife worlds and the choice of the ordering of these worlds (see Eells, 1983). Since the frequency theory provides no non-arbitrary basis for making these choices, the proposed truth conditions seem to rely upon a measure theoretic notion not explicable in frequency terms. In the following chapter we will see one possible basis for this measure.

The hypothetical frequency approach is not only unworkable but is wrong-headed as well. Suppose that some appropriate explication of 'almost all' has been given and that almost all of the infinite repetitions of the chance experiment result in a limiting frequency of p for '0'. Would we then be justified in concluding that the probability of '0' is p? Not if the outcome sequences lack other probability 1 properties characteristic of the given p-value. Which probability 1 properties shall we then take to be definitory of probability? One solution, suggested by von Mises' work, would focus on those properties characteristic of randomness for the given p-value. We have seen, however, that there are many different possible explications of randomness for infinite sequences. The frequency theorist must hold that only one of these possible explications captures the true notion of randomness, or else he must admit that there are many different concepts of physical probability, one for each possible definition of randomness. The need to divine the 'true' concept randomness or else to multiply concepts of probability is avoided if we reject the need for a 'definition' of probability in terms of relative frequency or the like. Although they are not explicit on this point, frequency theorists seem to have two motivations, both bad, in striving for a definition. First, there is the felt need to establish a connection between probability assertions and observable and measurable quantities such as frequencies. But the connection doesn't require the addition of a new axiom or definition to probability theory; standard probability theory already establishes whatever valid connection exists between probabilities and limiting frequencies via the strong law of large numbers (compare Doob (1941) with von Mises (1941)). Second, there is the related desire to locate physical probabilities within the Empiricist framework. But again, a definition of probability as hypothetical limiting frequency is not a requisite condition for this goal.

The example of ergodic theory studied in the next chapter provides a *tertium quid* to propensity and frequency theories. It eschews the definitional approach of the frequency theory but shows how the physical probabilities of classical statistical mechanics can be grounded in accord with the Empiricist demands. The example is especially interesting from the point of view of this work since the laws that do the grounding are deterministic.

The inference: 'objective' as applied to probabilities implies that the probabilities are irreducible, which in turn implies that determinism is false: goes wrong at the first step. $p = \frac{1}{2}$ (say) for a coin can represent an objective tendency for the coin to land heads up. And the attribution of such an objective tendency is not undercut by the discovery that the outcome of any flip is uniquely determined by the prior micro-state of the system; indeed, if classical statistical mechanics is our guide, determinism can form part of the explanation of the tendency to land heads up $\frac{1}{2}$ of the time. The second step of the inference can be secured by making it part of the meaning of 'irreducible'. Then to imagine a world with irreducible probabilities is necessarily to imagine an indeterministic world. If we try to imagine a world whose laws are not only indeterministic but also purely stochastic we imagine ourselves into a problem for the grounding of probabilities. Certainly the frequency theorist's attempt at a definitional grounding is of no help. Recall that the hypothetical frequency approach bases the truth of a probability assertion on the relative frequencies in hypothetical situations obeying the laws of the actual world. On pain of circularity, 'laws' here must be taken to mean non-probabilistic laws; but since in our imagined stochastic world there are no such laws the frequentist truth conditions collapse. Since we have rejected the frequency theory this collapse need not be seen as an immovable obstacle to an Empiricist grounding of physical probabilities. But until it is shown how such a grounding can be provided in non-deterministic worlds, we are shadowed by a dilemma: with determinism irreducible probabilities are impossible; without determinism irreducible probabilities are possible but they portend non-Empiricist powers.

NOTES

[1] Perhaps we should put scare quotes around 'laws' since at this stage we do not know whether there are in nature any irreducibly probabilistic laws or even what it would mean to say that there are such laws.

[2] This is needed for the strong form of the law of large numbers used later.

[3] Schnorr (1971, 1971a) has provided a general framework for discussing effective statistical tests for infinite sequences. An effective test is defined informally in terms of a function $F: X_0^* \to \mathbb{R}$ such that

(A1) F is given by algorithms
(A2) There is rule which assigns to F a null set $\mathcal{N}_F$, the set of sequences that fail the test.

Given a precise implementation of these axioms, we can say that an infinite sequence is random just in case it passes all effective tests. To carry forward the gambling analogy, we could think of $F(x(n))$ as the again after the nth trial. Then for a fair coin ($p = \frac{1}{2}$) we would want the Martingale property that $F(x) = \frac{1}{2}F(x0) + \frac{1}{2}F(x1)$. The associated null set could be defined as $\mathcal{N}_F = \{x \in X_0: \limsup F(x(n)) = \infty\}$. However, Schnorr recommends that we exclude the possibility that $F(x(n))$ grows so slowly that the growth cannot be recognized by effective means. Write $\mathrm{elim\,sup}\, F(x(n)) = \infty$ if there is recursive monotone unbounded $g: \mathbb{N} \to \mathbb{N}$ such that $\limsup (F(x(n)) - g(n)) \geq 0$. Then Schnorr's associated null set is $\{x \in X_0: \mathrm{elim\,sup}\, F(x(n)) = \infty\}$.

[4] It might seem more natural to write "for every n" in place of "infinitely often as $n \to \infty$." We will see in Sec. 3 that this is not a live option.

[5] If the test is to be effective, the p-value must be a computable number; see Martin-Löf (1966).

[6] Alternatively, we could say that a finite sequence is $(\varepsilon, n, \mathcal{F})$ random with respect to a p-value if the frequency of 0s in the sequence is approximately (within ε) p and is approximately (again within ε) invariant under any selection rule which is in the family $\mathcal{F}$ and which selects out a subsequence of length at least as great as n (see T. Fine (1973)). For infinite sequences the $(\varepsilon, n, \mathcal{F})$ relativization can be removed by taking ε to be 0, n to be $+\infty$, and $\mathcal{F}$ to be the family of all effective selection rules. For finite sequences the relativization to ε and n cannot be removed, and the removal of the relativization to $\mathcal{F}$ is problematic. Finite sequences cannot stand up to all effective selection rules, and the limitation to 'non-contrived' effective rules puts randomness at the mercy of our intuitions of naturalness.

[7] What follows is a version of a proof that was kindly provided by D. Malament and S. Zabell.

[8] Proponents of the propensity theory include Popper (1959, 1962); Giere (1973, 1976); and Fetzer (1981). Frequency theorists include von Mises (1941, 1957, 1964); Reichenbach (1971); and van Fraassen (1977, 1980). The survey articles by Kyburg (1974) and Eells (1983) provide good overviews.

[9] See Kyburg (1974). I have made no attempt to follow the niceties of Kyburg's definitions. And to simplify the discussion I ignore the possibility that the laws of physics might make impossible an infinite repetition of the chance experiment by implying, for example, that the world ends after a finite time, as can happen in general relativistic cosmological models (see Ch. X).

SUGGESTED READINGS FOR CHAPTER VIII

An accessible treatment of von Mises' theory of Kollektives and the von Mises-Church definition of random sequence is to be found in Martin-Löf's (1969) "The Literature on von Mises' Kollektives Revisited." Kyburg's (1974) "Propensities and Probabilities" describes various frequency and propensity theories of probability. T. Fine's (1973) *Theories of Probability* is a gold mine of information on many of the topics of this chapter. Chatin's (1975) *Scientific American* article "Randomness and Mathematical Proof," gives a sketch of the computational complexity approach to randomness and also links it to Gödel's incompleteness theorem.

CHAPTER IX

DETERMINISM, INSTABILITY, AND
APPARENT RANDOMNESS

> ... the rock loosed by frost and balanced on a singular
> point of the mountainside, the little spark which
> kindles the great forest, the little word which sets the
> world fighting, the little scruple which prevents a man
> from doing his will, the little spore which blights all
> the potatoes, the little gemmule which makes us
> philosophers or idiots ... At these points, influences
> whose physical magnitude is too small to be taken
> account of by a finite being, may produce results of the
> greatest importance.
> (James Clerk Maxwell, "Does the progress of Physical
> Science tend to give any advantage to the opinion of
> Necessity (or Determinism) over that of the Con-
> tingency of Events and the Freedom of the Will?")

In the terminology of Hadamard's classic work, *Lectures on Cauchy's Problem*, a Cauchy initial value problem is "correctly set" if the initial data determine a unique solution *and* solutions depend continuously on the initial data. Implicit in this terminology is a methodological injunction: if your mathematical model of a physical system entails an initial value problem which exhibits non-uniqueness or else uniqueness but instability, then assume that the fault is not in Nature but in your model. In previous chapters we have seen how fruitful the first part of this injunction can be; tracing the reasons for a failure of uniqueness often leads to the discovery of restrictions on 'physical' solutions that were neglected in the original model. The second part of the injunction has also proved fruitful but to a lesser degree. For despite the seemingly straightforward appeal of the notion of continuous dependence of solutions on initial data, there are many inequivalent ways to gauge continuity, and especially in the case of partial differential equations, the verdict on stability can vary with the choice of gauge. For particles as opposed to fields, the gauge is much less open to choice, but we have learned that in the case of particles Nature sets and solves many 'non-correctly set' problems where determinism but not stability holds. Moreover, this instability is the foundation of one of the important

154

bridges from micro-determinism to macro-randomness. The main purpose of this chapter is to show that by crossing this bridge a microdeterministic system can on the macro-level violate even the weak form of determinism studied in Ch. VIII and behave like a Bernoulli system.

Though precise mathematical results in this area are of fairly recent origin the basic ideas have a venerable history. Writing in the 1870's, Clerk Maxwell was careful to distinguish the maxim that 'The same causes will always produce the same effects' from a second maxim that 'Like causes produce like effects'. The former he interpreted in terms of a time translation invariant determinism while the latter "is only true when small variations in the initial circumstances produce only small variations in the final state" (1920, p. 13). And while the latter demand is often satisfied, it is not always so, "as when the displacement of the 'points' causes a railway train to run into another instead of keeping its proper course." To this homely example Maxwell added the parenthetical but prophetic remark that

We may perhaps say that the observable regularities of nature belong to statistical molecular phenomena which have settled into permanent stable conditions. In so far as the weather may be due to an unlimited assemblage of local instabilities, it may not be amenable to a finite scheme of law at all. (1920, p. 14).

1. STABILITY AND INSTABILITY FOR FIELDS

Stability in the initial value problem for a field law can be formulated abstractly as the requirement that for any two sequences of solutions $\{u^n\}$, $\{u'^n\}$, convergence in the initial data implies convergence overall, i.e., if $(u^n(x, 0) - u'^n(x, 0)) \to 0$ as $n \to \infty$, then $(u^n(x, t) - u'^n(x, t)) \to 0$ as $n \to \infty$. Different senses of stability are produced depending upon how uniform the convergence is required to be and upon how the convergence is to be measured. If we have a norm $\| \ \|$ defined on instantaneous states $u(t) = u(\cdot, t)$, we can take future stability to mean that whenever $\|u^n(0) - u'^n(0)\| \to 0$, then also $\|u^n(t) - u'^n(t)\| \to 0$ for $t > 0$.[1] There are generally many choices for $\| \ \|$, but any acceptable choice must meet the demand that if u and u' agree on what is to be counted as initial data, then $\|u(0) - u'(0)\| = 0$. It then follows from stability that initial data determine a unique solution, and in fact determinism is frequently demonstrated as a corollary of a stability proof. When convergence of solutions is required to be uniform on compact intervals of t, we can rephrase stability as the requirement that there is non-negative and non-decreasing function $F(t)$ such that for any solutions u and u',

$$(IX.1) \quad \|u'(t) - u(t)\| \leqslant F(t) \|u'(0) - u(0)\|$$

We can illustrate these ideas for the examples of field laws studied in Chs. III and IV. For the shock wave equation (III.7) consider the space of weak solutions which are piecewise continuous with all discontinuities in the form of shocks, and let the norm be the L^1 norm. Then $\|u'(t) - u(t)\|$ is a decreasing function for $t > 0$ for any solutions u and u'. Future stability and determinism are immediate (see Lax (1973)).

For the heat equation (III.4) applied to a unit rod with temperatures at the ends kept at zero, consider the space of continuous solutions equipped with the sup norm or else the space of L^p ($p \geqslant 1$) solutions equipped with the L^p norm. Then for $t > 0$, $\|u(t)\| \leqslant \|u(0)\|$, and using linearity, future stability and determinism are immediate. However, past stability, which would require that the function $F(t)$ in (IX.1) exists for negative time and that $F(-t)$ is non-decreasing for $t > 0$, fails. In fact, there is such a great instability in the past direction of time that the slightest inaccuracy in the measurement of the final temperature distribution will undercut any finite retrodiction task of retrodicting with specified finite accuracy the temperature distribution over a specified finite past interval of time. In more ordinary $\varepsilon - \delta$ talk, for any $t < 0$, any $\varepsilon > 0$, and any $\delta > 0$, there are solutions u and u' such that $\|u'(0) - u(0)\| \leqslant \delta$ but $\|u'(t) - u(t)\| > \varepsilon$. (In the sup norm, consider $u \equiv 0$ and $u'_{\delta, n}(t) = \delta \sin(n\pi x)\exp(-n^2 t)$.)

For the relativistic wave equation (IV.1) consider first the ordinary C^2 solutions with finite energy. If we take the norm to be the energy norm $\| \ \|_E$ then past and future stability follow from linearity and the fact that $\|u(t)\|_E = \|u(0)\|_E$ for all t. However, if we want to consider weak $C_0(\mathbb{R}^3)$ solutions (continuous solutions which vanish at spatial infinity) the appropriate norm might seem to be the ordinary sup norm. But then stability is lost since focusing effects can cause a wave initially small to grow unboundedly large in the focused region. For Maxwell's equations, which reduce to the relativistic wave equation, stability also fails in the sup norm applied to electromagnetic fields whose components are $C_0(\mathbb{R}^3)$. It also fails in the L^p norm for solutions whose components are $L^p(\mathbb{R}^3)$ unless $p = 2$. There are ways to restore stability by renorming the space of solutions, but these matters are too technical to permit discussion here (see Fattorini (1983) for details).

2. CLASSICAL DYNAMICAL SYSTEMS

In the mathematician's terminology, an *abstract dynamical system* is a triple $(X, \{\phi_t\}, \mu)$. The space X is usually assumed to have a manifold structure. $\{\phi_t\}$ is a one-parameter family of automorphisms $\phi_t: X \to X$, $t \in \mathbb{R}$, with the group properties $\phi_0 = \text{id}$, $\phi_{t_2} \circ \phi_{t_1} = \phi_{t_1 + t_2}$, $\phi_t^{-1} = \phi_{-t}$. This is just a formalized way of reexpressing the assumptions of futuristic and historical determinism in time translation invariant laws for a classical system: the possible instantaneous states of the system correspond one-one to the points of X and ϕ_t is the time evolution operator giving the state $\phi_t(x)$ at time t when the initial state at $t = 0$ is $x \in X$. The μ of the abstract dynamical system is a measure on X which is normalized ($\mu(X) = 1$) and which is invariant under the flow ($\mu(A) = \mu(\phi_{-t}(A))$ for any measurable $A \subseteq X$).

We are now in a position to state a version of *Poincaré's recurrence theorem*. Let $U \subseteq X$ be a measurable set and let U_0 denote the set of all $x_0 \in U$ such that there is a $T > 0$ such that for all $t \geq T$, $\phi_t(x_0) \cap U = \emptyset$, i.e., the orbit determined by x_0 eventually leaves U permanently. Then $\mu(U_0) = 0$.

In the Hamiltonian formulation of classical mechanics $X = \mathbb{R}^{2n}$ is called the phase space, and a phase point is denoted by $x = (q_1, q_2, \dots, q_n, p_1, p_2, \dots, p_n)$ where the q_i and p_i are respectively the canonical position and momentum coordinates satisfying Hamilton's equations

$$(\text{IX.2}) \qquad \frac{dq_i}{dt} = \frac{\partial h}{\partial p_i}, \qquad \frac{dp_i}{dt} = -\frac{\partial h}{\partial q_i},$$

where $h(q, p)$ is the Hamiltonian. The equations (IX.2) define a one-parameter group of automorphisms on $\mathbb{R}^{2n}$ called the Hamiltonian flow. The main application to be considered here is to particle mechanics; for N particles $n = 3N$ since three position coordinates and three momentum coordinates are needed to specify the instantaneous state of each particle. The Hamiltonian is a constant of the motion if, as we have assumed, h is not an explicit function of time:

$$\frac{dh}{dt} = \cancel{\frac{\partial h}{\partial t}}^{0} + \sum_i \left(\frac{\partial h}{\partial q_i} \frac{dq_i}{dt} + \frac{\partial h}{\partial p_i} \frac{dp_i}{dt} \right)$$

$$= \sum_i \left(-\frac{dp_i}{dt} \frac{dq_i}{dt} + \frac{dq_i}{dt} \frac{dp_i}{dt} \right) = 0.$$

The natural volume element $dq_1 \ldots dq_n dp_1 \ldots dp_n$ on $\mathbb{R}^{2n}$ is preserved by the Hamiltonian flow (*Liouville's theorem*) as can be shown by verifying that the divergence of the vector field $(dq_1/dt, \ldots, dp_1/dt, \ldots)$ vanishes. For a system confined to a finite volume of phase space we thus have a normed invariant measure and can apply Poincaré's recurrence theorem. Define a phase point x to be *recurrent* just in case for each open neighborhood $N(x)$ of x and every time $T > 0$ there is a $t \geqslant T$ such that $\phi_t(x) \cap N(x) \neq \emptyset$. Then the set of non-recurrent points has zero volume in phase space. Define a point x to be *quasi-recurrent* just in case points arbitrarily close to x eventually return to the vicinity of x, i.e., for any neighborhood $N(x)$ and any $T > 0$, there is a time $t \geqslant T$ such that $\phi_t(N) \cap N \neq \emptyset$. Then since the set of non-quasi-recurrent points is open and has zero volume, it is empty, i.e., every point is quasi-recurrent.

From here on I will assume that the system of interest is isolated and that all possible histories have the same constant energy E. Thus the Hamiltonian flow is confined to a $(2n - 1)$-dimensional energy surface S_E. The volume element on $\mathbb{R}^{2n}$ induces on S_E an invariant area measure dv, and if we assume that our system is spatially bounded (e.g., particles in a box with rigid walls), then $\int_{S_E} dv$ is finite and we can satisfy the definition of a dynamical system by taking $\mu(A) = \int_A dv / \int_{S_E} dv$. The physical justification for taking this measure as a guide to the probabilities of outcomes of macroscopic measurements will be discussed in due course.

3. MACRO-RANDOMNESS: WHAT WE WANT

For macro-randomness the first thing we want to be assured of is that the macro-past history of the system does not determine the future. To make this more precise, consider a partition of the time axis into finite intervals. $\tau > 0$, so that $t_{i+1} - t_i = \tau$, and a partition of the energy surface S_E into finite sized cells R_j ($\cup_j R_j = S_E$ and $R_j \cap R_k = \emptyset$ if $j \neq k$). Physically, we suppose that there is a macro-operation that takes time τ to perform, that allows us to tell into which of the cells R_j the state of the system falls, but that allows no finer discrimination. Then for these partitionings, the macro past history will be a record

$$\ldots R_{i_n}, R_{i_{n-1}}, \ldots, R_{i_1}$$

of the cells R_{i_k} into which the state was found to fall at times t_{-k}. To assert that the system behaves non-deterministically at the macro-level is to assert that for any such partitions of the energy surface and the time axis, the resulting past macro-history does not confer certainty on the macro-state at the next instant. Expressed in terms of the μ-measure this means that

(IX.3) $\mu(R_j/\phi_\tau(R_{i_1}) \cap \phi_{2\tau}(R_{i_2}) \cap \ldots \cap \phi_{n\tau}(R_{i_n}))$

does not approach 1 or 0 as $n \to \infty$ where $\mu(\cdot/\cdot)$ is the conditional measure defined by $\mu(A/B) \equiv \mu(A \cap B)/\mu(B)$. Interpreting μ as probability (an interpretation still to be justified), $\mu(R_j/\phi_\tau(R_{i_1}) \cap \phi_{2\tau}(R_{i_2}))$, for instance, is the probability that the state will be found in cell R_j at t_0 given that it was found in cell R_{i_2} at t_{-2} and in cell R_{i_1} at t_{-1}.

For macro-randomness we want more than lack of determinism. We want the system to behave as if it were a Bernoulli process. This requires that not only does the expression (IX.3) not approach certainty but that, as in the spin of a fair roulette wheel, the past history is probabilistically irrelevant to the future:

(IX.4) $\mu(R_j/\phi_\tau(R_{i_1}) \cap \phi_{2\tau}(R_{i_2}) \cap \ldots \cap \phi_{n\tau}(R_{i_n})) = \mu(R_j)$

for any j and any n. To make contact with the discussion in the preceding chapter, we can specialize to the case of a two-element partition $\{R_0, R_1\}$, coding R_0 with '0' and R_1 with '1'. Thus, a macro-history is a doubly infinite binary sequence with the probability of $p = \mu(R_0)$ of being '0' and a probability of $1 - p = \mu(R_1)$ of being '1'. The Bernoulli measure on the collection of all possible macro-histories (not to be confused with the μ of the dynamical system) is constructed as before.

These then are the ultimate properties of randomness we want our system to have. They are based on weaker properties which deserve at least honorable mention. The historically first and the weakest of the properties, called *ergodicity*, was introduced by Ludwig Boltzmann and Clerk Maxwell. On Maxwell's formulation, ergodicity meant that "the system, if left to itself in its actual state of motion, will, sooner or later pass through every phase consistent with the equation of energy [i.e., through every $x \in S_E$]".[2] We now know from dimensionality considerations that this original form of the ergodic hypothesis is, with one exception, false: since every orbit $\phi_t(x)$ is one-dimensional it cannot fill up S_E if $\dim(S_E) > 1$. However, we can preserve the spirit of Maxwell's idea in the form of the *quasi-ergodic hypothesis*: almost every $x \in S_E$

gives rise to an orbit $\phi_t(x)$ which meets every region $R \subset S_E$ such that $\mu(R) > 0$. (Here 'almost every' means except for a set of μ-measure 0.) Thus for an ergodic system the phase orbit wanders 'all over' the energy surface never becoming trapped in any proper sub-region: if $\phi_t(A) = A$ for all t, then $\mu(A) = 1$ or 0. Further, if we start with at any instant, say $t = 0$, and keep track of the relative sojourn time$(A)/t$ of the phase point in A, then in the very long run, $\lim_{t \to \pm\infty} \text{time}(A)/t = \mu(A)$.[3] The wandering nature of the orbits and the emergence of limiting frequency probabilities is some sign that the system behaves randomly at the macro-level. But the sign may not be strong enough. A simple harmonic oscillator is ergodic as well as quasi-ergodic, but if we choose τ to coincide with the period of the oscillator, $\phi_{n\tau}(R_j) = R_j$ and the desired properties of randomness are violated.

Another interesting feature of quasi-ergodicity is that it implies the non-existence of any other measure μ' which is invariant under the flow and absolutely continuous with respect to μ (i.e., if $\mu(A) = 0$ then $\mu'(A) = 0$).[4] This provides some justification for using μ in the abstract dynamical scheme since sets of zero μ-measure are intuitively negligible; but as yet we have no justification for aligning our expectations about the outcomes of macro-measurements to fit the μ-rule of equal probabilities to equal areas of S_E. Mixing is the next higher ergodic property which moves us in the direction of the required justification.

Formally, *mixing* demands that

$$(IX.5) \quad \lim_{t \to \pm\infty} \mu(\phi_t(A) \cap B) = \mu(A) \cdot \mu(B), \text{ for all measurable } A, B$$

A mixing system is necessarily quasi-ergodic. For if $\phi_t(A) = A$ for all t, we have from (IX.5), taking $B = A$, $\mu(A) = (\mu(A))^2$, implying that $\mu(A) = 0$ or 1. But the implication does not always go in the other direction — the simple harmonic oscillator is ergodic but not mixing.

To see how mixing connects with macro-expectations, suppose that we start with an ensemble of systems distributed over region A at time $t = 0$ with a constant density $\rho_0(x) = 1/\mu(A)$ if $x \in A$, 0 otherwise.

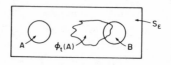

Fig. IX.1

Then by Liouville's theorem for Hamiltonian systems, $\rho_t(x) = \rho_0(\phi_{-t}(x))$, and so the fraction of points of the original ensemble which are in B at $t > 0$ is

$$\int_{\phi_t(A) \cap B} \rho_t \, dv = \int_{\phi_t(A) \cap B} \rho_0(\phi_{-t}(x)) \, dv = \frac{\mu(\phi_t(A) \cap B)}{\mu(A)}$$

For a mixing system this fraction approaches $\mu(B)$ for large enough t.

Applying this result to our macro-experiment, suppose that our observation at $t = 0$ places the state in cell R_k of the partition. Since our measurement does not allow any finer discrimination, our initial knowledge is represented by a uniform probability density over R_k at $t = 0$. If the system is mixing and our next observation is made at $t = \tau$ with large enough τ, then our conditional probability that the outcome will correspond to cell R_l, given the previous outcome, will be nearly $\mu(R_l)$.

The rub comes in the qualification 'large enough'; mixing (IX.5) is an asymptotic property and does not by itself assure that the mixing takes place within the chosen macro time scale. The Bernoulli property guarantees rapid mixing, for it implies that $\mu(R_l / \phi_\tau(R_k)) = \mu(R_l)$ so that $\mu(R_l \cap \phi_\tau(R_k)) = \mu(R_l) \cdot \mu(\phi_\tau(R_k)) = \mu(R_l) \cdot \mu(R_k)$. Thus, the Bernoulli property provides both the justification for taking μ as a guide to macro-expectations and the justification for believing that these expectations are equivalent to those for a roulette wheel.

4. MACRO-RANDOMNESS: CAN WE HAVE WHAT WE WANT?

We have reviewed what we need to have macro-randomness emerge from micro-determinism in classical dynamical systems. The question is now whether we can have what we want for physically realistic systems.

Integrable Hamiltonian systems — those possessing constants of motion other than the energy — cannot exhibit even the humblest of the ergodic properties; for the extra constants mean that a measurable set of states will generate orbits that cannot wander all over S_E in the manner required by quasi-ergodicity. For some time it was the fond hope of workers in classical statistical mechanics that the hierarchy of ergodic properties would be exhibited in typical non-integrable Hamiltonian systems. This hope was dashed by the work of Kolmogorov, Arnold, and Moser who proved that a wide class of non-integrable systems possess 'invariant tori' of dim n; so except in the case of $n = 1$ and $\dim(S_E) = 2n - 1 = 1$, not even the lowest order ergodic property can hold.[5]

We can look for randomness in the gaps between the KAM tori. Or we can try to escape the KAM result in one of two ways. The first is to go to a 'thermodynamic limit', in which the number N of particles goes to infinity while the ratio of N to the volume remains finite, hoping that the limit will force out the ergodic hierarchy. To even begin to formulate ergodic theory in the limit we would first have to prove existence and uniqueness theorems for the motions of infinite collections of particles — no easy task as we saw in Ch. III — and we would have to show how to transfer the measure theory to an infinite dimensional state space — again, a non-trivial task. The second tack is to leave N finite but to search for systems which do not fall prey to KAM. This is the line I will explore, not because it is necessarily the 'right' one but because it illustrates how it is possible for randomness to emerge from determinism even for small N; indeed, even $N = 1$ will do!

If we want to find a system which is sufficiently far up the ergodic hierarchy it is intuitively evident that we must search for an unstable system: the dispersal of points required by mixing and the higher order ergodic properities indicates that initially close phase points diverge rapidly.

5. INSTABILITY IN CLASSICAL PARTICLE SYSTEMS

For Hamiltonian mechanics the discussion in Sec. 1 of stability and instability for field theories can be paralleled by using a norm on the tangent vectors of the phase space. However, it is more intuitive to work with a measure of the distance between phase orbits. Any Riemann metric h on the phase space X will generate a distance measure $\text{dist}_h(x, x')$ defined as the greatest lower bound on the h-length of paths joining x and x'. When X is compact (as we nave been assuming here) the choice of h has no significant influence on conclusions about stability and so the subscript on dist will be dropped.

Using dist any number of different definitions of stability and instability can be concocted. For instance, *Lyapunov instability* expresses the failure of uniform convergence of phase orbits with respect to the initial data. The phase space is Lyapunov stable at x (or equivalently, Lyapunov stable with respect to the unique phase orbit $\phi_t(x)$ through x)

just in case for any $\varepsilon > 0$ there is a $\delta > 0$ such that for any y and any time $t > 0$, if dist$(y, x) \leqslant \delta$ then dist$(\phi_t(y), \phi_t(x)) \leqslant \varepsilon$. If the actual state of the system corresponds to a Lyapunov unstable point then for some $\varepsilon > 0$ the ongoing prediction task of forecasting the future state with accuracy ε for the indefinite future will prove to be impossible if there is even the slightest error in ascertaining the initial conditions. Other senses of stability can be related to the conditions needed to carry out various of Popper's other 'prediction tasks' (see Ch. II) in the face of uncertain knowledge of the initial positions and velocities of the particles. Our aim, however, is not to do Sir Karl's work but to investigate the kind of instability needed to generate stochastic behavior at the macro-level.

Note that, paradoxically, the framework we have chosen automatically entails that the systems under study cannot be radically unstable. The system has, by assumption, only a finite volume of phase space available to it, and as a consequence there will be stability in Poincaré's sense at almost all $x \in S_E$ in that $\phi_t(x)$ returns infinitely often to any chosen neighborhood $N(x)$ of x. This form of stability holds for all finite Hamiltonian systems and is not negated by ergodic properties. It is often pointed out that in typical macro-systems the magnitude of the Poincaré recurrence time is quite large — often larger than the age of the universe. But for our purposes this is irrelevant; for even if the recurrence time is short in comparison with the time scale of the macro-measurements, the system can nevertheless exhibit macrorandomness due to the facts that no macro-measurement can precisely determine the initial state and that phase points within the range of error can be rapidly dispersed and can have very different recurrence times. Here we meet a second way in which our framework limits instability. In dynamical systems the points in $N(x)$, $\mu(N(x)) < 1$, cannot disperse to cover the entire energy surface or even a larger area of it. And for a Hamiltonian flow it cannot be the case that all of the phase orbits emanating from points in $N(x)$ are dispersing (a form of Liouville's theorem).

We can, however, have Hamiltonian dynamical systems which are what Anosov calls the C-systems,[6] possessing a strong dispersal property. At each $x \in X$ there is a submanifold $X^d(x)$ of exponentially dilating orbits: for $x' \in X^d(x)$

$$\text{dist}(\phi_t(x), \phi_t(x')) \geqslant a \exp(\alpha t) \, \text{dist}(x, x'), \, t > 0$$
$$\text{dist}(\phi_t(x), \phi_t(x')) \leqslant b \exp(-\alpha t) \, \text{dist}(x, x'), \, t < 0$$

where a, b, and α are positive constants and $\dim(X^d(x)) > 0$ and is independent of x; there is also at each point $x \in X$ a submanifold $X^c(x)$ of contracting stable orbits defined by reversing the inequalities in the definition of $X^d(x)$.

Measure preserving flows with the C-property are, with one class of exceptions, *K-systems* whose technical definition I will not discuss except to say that K-systems have the first of the desired properties of macro-randomness; namely, the past macro-history does not determine the future macro-development. Concrete examples of Hamiltonian particle systems which have been proven, because of Anosov instability, to be K-systems (almost everywhere) are (1) two or more hard spheres in a rectangular box with perfectly reflecting walls and (2) a single point mass in a 'Born box' which contains stationary convex scatterers. Small changes in initial directions are rapidly multiplied as the spheres of model (1) collide with one another or the particle of model (2) collides with the stationary scatterers. Born's model is sufficiently unstable to generate a Bernoulli flow (almost everywhere): for any choice of $\tau > 0$ there is a partition of S_E coarse enough to have the Bernoulli property (IX.4) but fine enough that different dynamical states can be discriminated by some possible sequence of macro-observations.

6. STRANGE ATTRACTORS

Our discussion has neglected dissipative systems. At first glance such systems appear to be poor candidates for exhibiting random or chaotic behavior. Assuming futuristic determinism we still have a flow ϕ_t on the phase space of the system, but the flow may only have a semi-group property if ϕ_t is not invertible because of the loss of historical determinism. Further, neither energy nor volume in phase space need be preserved, and typically the phase space will contain regions A ('attractors') towards which the flow takes all the states in a neighborhood of A. (More formally, call $A \subset X$ an *attractor* under the flow ϕ_t if A is compact, $\phi_t(A) \subseteq A$ for all $t > 0$, and there is a neighborhood $N(A)$ of A such that all the future oribits originating in $N(A)$ are pulled into A, i.e., $\lim_{t \to +\infty} \phi_t(N(A)) \subseteq A$. It is usually also required that A be *indecomposable*, i.e., if $B \subseteq A$ is an open set and $\phi_t(B) = B$ for all $t > 0$, then either $B = A$ or B is empty.) In the simplest case the system dissipates some 'noble' form of energy, generating heat and settling down to an equilibrium state; and it settles down without the fuss or bother that is read on the macro-level as randomness or chaos. Nevertheless, if the attractors have a complicated enough structure and the flow is unstable enough ('strange attractors') there can be plenty of fuss or bother. (A *strange attractor* A under the flow ϕ_t is generally defined to be an attractor such that there is future Lyapunov instability at every point in some neighborhood of A.)

An indication of the fuss and bother can be supplied by a technique first exploited by Poincaré. Choose an $(n-1)$-dimensional section S of the phase space X and keep track of the successive intersections with S of an orbit under ϕ_t. With strange attractors present the pattern of intersections can look quite irregular. Poincaré's device also provides an alternative way to characterize strange attractors. By futuristic determinism, if the future orbit through $x \in S$ meets S again the next meeting place $y \in S$ is unique; $f(x) = y$ then defines the Poincaré map on S or on the subset of all the points of S which generate reintersecting orbits (usually S can be chosen so that f is defined on all or almost all of S). An attractor $A \subset S$ of S is then defined as before using iterates of the Poincaré map in place of ϕ_t. (The attracting properties of A are that $f(A) \subseteq A$ and that there is a neighborhood $N(A) \subset S$ such that $\lim_{n \to \infty} f^n(N(A)) \subseteq A$, and Lyapunov instability at $x \in S$ means that there is an $\varepsilon > 0$ such that for any $\delta > 0$ there is a $y \in S$ and an $n > 0$ such that dist$(y, x) \leqslant \delta$ but dist$(f^n(y), f^n(x)) > \varepsilon$.)

In 1963 Edward Lorenz modeled the process of atmospheric convecting using the system of equations

$$(\text{IX.6}) \quad \frac{dx}{dt} = -ax + ay$$

$$\frac{dy}{dt} = -xz + bx - y$$

$$\frac{dz}{dt} = xy - cz$$

Oscar Lanford has charted the behavior for the case $a = 10$, $b = 28$, and $c = 8/3$. A solution which starts at the origin $(0, 0, 0)$ of the phase space $\mathbb{R}^3$ makes a loop to the right around $(6\sqrt{2}, 6\sqrt{2}, 27)$, then several loops to the left around $(-6\sqrt{2}, -6\sqrt{2}, 27)$, then more loops to the right, etc. in an irregular pattern (see Fig. 3 of Ruelle (1980)). Sensitivity to initial conditions is indicated by the fact that the slightest change in the initial state changes the number of loops to the left and to the right. Little wonder, if Lorenz's model is correct, that the weather man has such a tough time forecasting! It can be shown that all of the future orbits in the phase space are pulled into a bounded region B of $\mathbb{R}^3$. Further, the flow ϕ_t shrinks volume at the rapid rate $\exp(-13.67t)$ so that in unit time a unit volume has shrunk to roughly $1/10^6$. Thus, all of the future orbits converge to $\lim_{t \to +\infty} \phi_t(B)$, which has measure zero. (In other cases, the attractors can have a positive invariant measure.) For the chosen values of the parameters there are stationary solutions at $(6\sqrt{2}, 6\sqrt{2}, 27)$ and $(-6\sqrt{2}, -6\sqrt{2}, 27)$, and for the section $z = 27$ the Poincaré map is defined almost everywhere; for a description of the fascinating and intricate details of the action of this map the reader is referred to Ruelle (1977a) and Lanford (1977). Computer calculations indicate that the time correlation between functions of state approaches zero as $t \to +\infty$ — a feature which in Hamiltonian systems is equivalent to the property of mixing — and that this mixing-like behavior occurs at a rapid and, possibly, exponential rate.

It is hoped that the study of strange attractors will have payoff in the form of an explanation of turbulence in fluids and gases, one of the most familiar but least understood phenomena of physics. The older approach of Landau and Lifshitz (1959), which defined turbulence in terms of quasi-periodic orbits, was found to be defective in various ways: it does not always lead to mixing-like behavior and it does not apply to those viscous fluids which may not possess quasi-periodic orbits.

7. CONCLUSION

No attempt was made to answer the still controversial question of what in fact accounts for the apparent macro-randomness in classical dynamical systems. Perhaps it is due to internal instability; or to instability under external perturbations; or to a thermodynamic limit; or to a fourth factor; or to a combination of factors. The points to be made were conceptual: determinism need not march in lock step with the stronger property of stability in the initial value problem; and when stability and determinism are out of step, prediction and determinism can part company if the predictions are to be based on initial data that involve even the slightest error; and more, certain kinds of unstable micro-determinism are not only compatible with but actually entail macro-randomness. Because of such cases we can anticipate one of the lines of defense a determinist will use to resist the move from apparent randomness to the conclusion of an ultimate, irreducible randomness in nature. Whether or not such a defense is unbreachable is a question we will have to confront in Ch. XI on quantum physics.

APPENDIX: THE BAKER'S TRANSFORMATION

The 'baker's transformation' provides the simplest known explicit example of a measure preserving flow with the Bernoulli property. The flow is non-Hamiltonian and it operates in discrete time; nevertheless, it is useful in illustrating some of the features we expect to find in Hamiltonian systems. Most of the standard works on ergodic theory discuss this example, but because of the rather forbidding character of these references, it may not be amiss to outline some of the essential features here.

In a way, the example is a cheat: we start with a Bernoulli process and then work backwards toward what we want. Recall the treatment of a binary Bernoulli process (e.g., coin flipping) from Ch. VIII. Assume that the probabilities for 'heads' (0) and 'tails' (1) are each 1/2 and that the flips are probabilistically independent. We indicated how to use these probabilities to construct a Bernoulli measure μ on X^∞, the possible outputs of the process (i.e., doubly infinite binary sequences). The new element added here is the Bernoulli shift transformation ϕ: $X^\infty \to X^\infty$, where $\phi(x) = x'$ with $x'_k = x_{k+1}$. It is obvious that ϕ is measure preserving: for example, if X^i_k is the cylinder set $\{x \in X^\infty : x_k = i\}$, $i = 1$ or 0, then $\mu(\phi(X^i_k)) = \mu(X^i_{k+1}) = \frac{1}{2} = \mu(X^i_k)$.

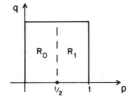

Fig. IX.2

The trick is to transfer this apparatus to a 2-dimensional 'phase space' $\mathscr{P}$ whose points (q, p) fall in the unit square $0 \leqslant q, p \leqslant 1$. Each binary sequence $x \in X^{\infty}$ corresponds to a phase point

$$\text{(IX.7)} \quad q = \sum_{k=1}^{+\infty} \frac{x_k}{2^k}, \qquad p = \sum_{k=0}^{-\infty} \frac{x_k}{2^{-k+1}}$$

Writing (q, p) in binary notation $(0.q_1q_2 \ldots, 0.p_1p_2 \ldots)$, the action of ϕ is: $\phi(0.q_1q_2 \ldots, 0.p_1p_2 \ldots) = (0.p_1q_1q_2 \ldots, 0.p_2p_3 \ldots)$. Or in more familiar notation

$$\text{(IX.8)} \quad \phi(q, p) = \begin{cases} (q/2, 2p) \text{ if } 0 \leqslant p < \tfrac{1}{2} \\ (q/2 + 1/2, 2p - 1) \text{ if } \tfrac{1}{2} \leqslant p \leqslant 1 \end{cases}$$

The name 'baker's transformation' comes from the fact that (IX.8) is like the kneading of dough; it squashes the loaf down, tears it in half, and stacks one half on top of the other. The ϕ of IX.8 provides the dynamics of $\mathscr{P}$: if at time $t = n$ the state of the system is (q, p), then at $t = n + 1$ the state is $\phi(q, p)$, and at $t = n + m$ it is $\phi^m(q, p)$. Since ϕ is invertible, we have both historical and futuristic determinism operating in discrete time. (Note that, considered by itself, the evolution of the p coordinate is futuristically but not historically deterministic.) The measure μ on X^{∞} transfers to the microcanonical measure on $\mathscr{P}$ giving equal weights to equal areas.

At the macro-level we have random behavior with respect to the two-fold partition $\{R_0, R_1\}$ of phase space given by R_0: $0 \leqslant p < \tfrac{1}{2}$, R_1: $\tfrac{1}{2} \leqslant p \leqslant 1$. We see that $(q, p) = (0.q_1q_2 \ldots, 0.p_1p_2 \ldots)$ is in R_0 or R_1 according as p_1 is 0 or 1. So from the construction of μ we have the desired properties of randomness, $\mu(R_0) = \mu(R_1) = \tfrac{1}{2}$ and $\mu(R_i/\phi(R_j)) = \mu(R_i)$. Nevertheless, if $(q, p) \neq (q', p')$, then there is an n such that $\phi^n(q, p) \in R_0$ and $\phi^n(q', p') \in R_1$ or vice versa.

NOTES

[1] Note that this sense of continuous dependence on initial data is not the same as Hadamard's (1952); see Fattorini (1983) for details.

[2] Maxwell (1890). For a history of the ergodic hypothesis, see Brush (1976).

[3] A theorem of Birkhoff shows that in any abstract dynamical system, ergodic or not, the time average exists for almost all orbits. Ergodicity is needed to show that the average is independent of the orbit and equal to $\mu(A)$.

[4] Malament and Zabell (1980) show that absolute continuity is equivalent to a requirement ("translation continuity") with a more obvious physical content. Translation continuity demands that the probability of finding the system in an open set varies continuously with small displacements of that set.

[5] See Arnold and Avez (1968) and Markus and Meyer (1974). However, a result of Oxtoby and Ulam (1941) goes somewhat in the other direction. They show that among the continuous measure preserving automorphisms of certain compact spaces the ergodic or metrically transitive ones are generic in the topological sense that they constitute all the automorphisms except for a set of first Baire category.

[6] See Arnold and Avez (1968). The more suggestive term 'Y-system' (from the Russian 'усы'-mustaches) is also used.

SUGGESTED READINGS FOR CHAPTER IX

A very readable survey of ergodic theory is to be found in the article "Modern Ergodic Theory" by Lebowitz and Penrose (1973). David Ruelle's (1980) "Strange Attractors" introduces this topic to non-specialists. Under the somewhat misleading title "Indeterminism in Classical Physics," Hoering (1969) provides a nice discussion of various concepts of instability. More advanced treatments of ergodic theory, turbulence, chaos, attractors, etc., include: Arnold and Avez (1968), *Ergodic Problems of Classical Mechanics*; Helleman (1980), "Self-generating chaotic behavior in non-linear mechanics"; Lichtenberg and Lieberman (1983), *Regular and Stochastic Motion*; and Garrido (1983), *Dynamical Systems and Chaos*.

CHAPTER X

DETERMINISM IN GENERAL RELATIVISTIC PHYSICS

> Of the general theory of relativity you will be convinced, once you have studied it. Therefore I am not going to defend it with a single word.
> (A. Einstein to A. Sommerfeld, 8 February 1916)

For Newton, "Absolute space, in its own nature, without relation to anything external, remains always similar and immovable." This is also an accurate description of the space-time background of both classical and special relativistic worlds. In general relativistic worlds, however, the space-time does not remain similar and immovable; it is rather an active participant in the unfolding drama of the world, and as a result its structure varies from physically possible world to physically possible world, and perhaps, from one instant to another within the same world. Since space-time is no longer a fixed canvas on which the world history is to be painted, the laws of nature, if they are to be deterministic, must specify how the structure of space-time itself evolves — the canvas and the painting are constructed simultaneously, a neat conjuring trick if it can be brought off.

1. GENERAL RELATIVISTIC WORLDS

A general relativistic space-time consists of a connected differentiable manifold M without boundary and Lorentz signature metric g defined on all of M. Minkowski space-time is included as the special case where M is the standard $\mathbb{R}^4$ and g is the Minkowski metric (see Ch. IV above). To serve as the basis of some space-time metric g the manifold M must either be non-compact or else, if compact, must have Euler characteristic zero. Otherwise, the topology of M can be quite wild. However, it will be assumed here that the topology is tame enough that the space-time is temporally orientable so that it admits a globally consistent time sense. (M, g is said to be *temporally orientable* just in case there is a continuous division of the lobes of the null cones of g into two classes, the 'past' and the 'future' lobes. Choose any point $p \in M$ and label one of the lobes of the null cone at p 'future' and the other 'past'. Then

170

choose any closed curve α through p, and transport a future pointing tangent vector at p around α by any method that is continuous and keeps timelike vectors timelike. Temporal orientability requires that upon return to p the vector shall not have flipped over into the past lobe of the null cone at p (see Fig. X.1). If M, g is not temporally orientable there always exists a covering space-time that is; thus, the failure of temporal orientability can be conceived as the result of the Creator's having made some nasty identifications of events.) How the assignment of the future direction of time is to be made is part of the problem of the direction of time, a problem which I will not attempt to resolve here.

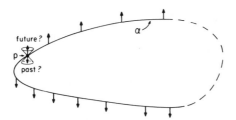

Fig. X.1 (Non-temporally orientable space-time)

The distribution of matter-energy throughout the space-time is described by a means of the stress-energy tensor T, a symmetric tensor field of type $(2, 0)$. The exact form of T must be derived from knowledge of the matter-energy fields present, but whatever the details of T, it is required to satisfy three structural principles: (i) Local conservation of energy in the sense that $T^{ij}_{\ \ ;j} = 0$. (ii) The energy condition, $T^{ij}V_i V_j \geqslant 0$ and $T^{ij}V_j$ non-spacelike for every timelike vector V^i. This condition implies that, as measured by any observer, the energy density is non-negative and the energy flow is non-spacelike.[1] (iii) T vanishes on an open neighborhood $U \subset M$ if and only if all matter fields vanish on U.

The Einstein field equations specify the relations between the structure of space-time, as given by g, and the distribution of matter-energy, as given by T, in terms of a set of 10 partial differential equations. Two equivalent forms of the equations are

$$(\text{X.1}) \quad R_{ij} - \tfrac{1}{2}R g_{ij} + \Lambda g_{ij} = 8\pi T_{ij}$$

and

$$(X.2) \quad R_{ij} = 8\pi(T_{ij} - \tfrac{1}{2}Tg_{ij}) + \Lambda g_{ij}$$

where R_{ij} is the Ricci tensor, R the scalar curvature, T = Trace (T), and Λ is the cosmological constant.[2] The conservation law (i) is an immediate consequence of (X.1)—(X.2).

A general relativistic *cosmological model* consists of a triple M, g, T satisfying the conditions listed above. If physically possible worlds correspond one-to-one with these models, then Laplacian determinism fails and fails miserably. The sections that follow investigate some of the more important hedges needed to make determinism viable in the large and the small.

2. TIME SLICES

Kurt Gödel (1949) found a remarkable solution to Einstein's field equations. Gödel's cosmological model is a dust filled universe ($T^{ij} = \rho V^i V^j$ where ρ is the density of the dust and V^i is the normalized velocity field of the dust). The dust is everywhere rotating relative to the local inertial structure; that seems a bit strange, but in itself it hardly prepares us for the mind boggling global features of the Gödel space-time. The Gödel manifold M is the standard $\mathbb{R}^4$. That implies that the space-time is temporally orientable, and in keeping with the stipulation of the preceding section, I assume that the time directionality has been fixed. The space-time trajectory of each dust speck is a timelike geodesic, and each such world line is open, i.e., topologically a real line. And yet, through each event in the space-time, there is a closed, future-directed, timelike curve. It follows that the Gödel model does not contain a single global time slice! Assume for purposes of contradiction that such a slice S exists. S would be two-sided, for by definition S is spacelike and the everywhere defined, continuous, and timelike vector field which gives the temporal orientation is non-tangent to S. Pick any point x on S. There is a future directed timelike curve which departs from S in the future direction from side 1 and returns to S from side 2. Such a curve cannot get around to side 2 by intersecting S from side 1, for then temporal orientability would be contradicted. Nor can it get to side 2 by going around an 'edge' of S since S is a global time slice. And finally, it cannot get to side 2 by travelling around a 'doughnut hole' in the space-time since the standard $\mathbb{R}^4$ does not have any such holes. We have run out of possibilities and into a contradiction.

There are three ways to construe the doctrine of Laplacian determinism in the large: as being conditioned on the existence of global
time slices; as unconditionally asserting their existence; or as pre-
supposing their existence as a condition of applicability of the doctrine.
The first alternative is unacceptable since it has the consequence that
Laplacian determinism in the large holds for any universe which admits
no time slice. We are left with the conclusion that either global
Laplacian determinism fails to apply to Gödel's universe or else applies
falsely. I will leave it to the reader to choose between these two ways of
describing the failure of determinism.

3. PARTITIONING BY TIME SLICES

Suppose that we restrict attention to cosmological models which can be
partitioned by time slices. Even within the bounds of this restriction,
determinism fails. As a trivial example, set the cosmological constant
$\Lambda = 0$ and consider two models, both of which are empty ($T = 0$
everywhere); one has Minkowski space-time as its space-time, the other
has a rolled up version obtained by identifying two points (x_1, t_1) and
(x_2, t_2) just in case $x_1 = x_2$ and $t_1 = t_2$ modulo some positive constant.
As illustrated in Fig. X.2, the two models agree on the slices S_a and S_b
(and on finite neighborhoods thereof) but do not agree globally.

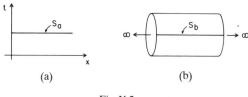

Fig. X.2.

4. CAUSALITY CONDITIONS

The examples from the two preceding sections involve intuitively
objectionable causal features. Thus, besides the question-begging desire
to save determinism, there appears to be an independent motivation for
narrowing the class of physically possible worlds as previously defined.
The first problem facing such a tack is an *embarras de richesses*: there
is a large hierarchy of ever stronger causality conditions to choose
from. The most attractive solution from the point of view of deter-

minism is to choose a condition strong enough to rule out all causal anomalies like closed, or almost closed, or almost . . . almost closed, causal curves but which is not so strong as to beg the question of determinism. Part of this bill seems to be filled by the requirement of *stable causality* which is satisfied by a space-time *M, g* just in case there is another space-time *M, g'* whose null cones are wider than those of *M, g* (i.e., at each *x* ∈ *M* every tangent vector which is non-spacelike with respect to *g* is likewise non-spacelike with respect to *g'* but not conversely) and which does not contain closed timelike curves. The physical motivation for stable causality is that we do not want a small perturbation in the mass-energy distribution and, hence, in the metric to eventuate in a closed causal loop.

Stable causality can be shown to be equivalent to the existence of a global time function in the sense of a differentiable map *t*: *M* → ℝ whose gradient is timelike (see Hawking and Ellis (1973)). The level surfaces *t* = constant then partition *M, g* by the time slices. Stable causality rules out every causal anomaly in the form of closed, almost closed, almost . . . almost closed causal curves and is thus arguably the strongest reasonable causality condition. Of course, there is a loose sense in which any condition framed in terms of the null cone structure can be counted as a causality condition; but if the condition does more than rule out causal anomalies in some sufficiently tight sense, it is in danger of begging the question of determinism.

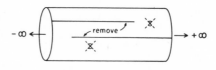

Fig. X.3

Having touted stable causality, I now must note that it guarantees neither that the world possesses an intuitively acceptable time structure nor that the world is made safe for determinism. By removing two baffles from the space-time of Fig. X.2(b) we get the stably causal space-time illustrated in Fig. X.3.[3] This example provides some intimation of how difficult a road we have before us in trying to make general relativistic worlds safe for determinism without blatantly legislating in favor of determinism.

5. THE STATUS OF CAUSALITY CONDITIONS

In evaluating the proposed imposition of a causality condition — whether it be stable causality or some other — two kinds of cases should be distinguished. In the case of the actual world, we do have some evidence against the existence of a space-time structure which permits causal anomalies. The existence of closed timelike curves, for example, usually imposes consistency conditions on relativistic equations of motion or field equations of the usual hyperbolic type. As an example of how severe the consistency conditions can be, Hawking and Penrose (1970) found that in the acausal toroidial space-time formed from two-dimensional Minkowski space-time by identifying the points (x, t) and $(x + m, t + n\pi)$, n, m positive integers, the only solution to the scalar wave equation $\partial^2 u/\partial t^2 - \partial^2 u/\partial x^2 = 0$ is $u = $ constant. But in the portion of the universe we can observe, there is an impressive variety of initial and boundary conditions, and moreover, the only restrictions on our ability to create new initial and boundary conditions seem to be those imposed either by known laws or engineering limitations. Of course, such arguments need to be handled with care. In the finite number of observations we make, we never see enough variety to rule out all possible cases of closed causal curves passing through the portion of the universe observationally accessible to us. Or perhaps the consistency conditions imposed by acausal space-time structure operate in some subtle manner on our desires so as to assure that we never try to set up configurations that would violate these conditions. Nor does the existence of closed causal curves invariably give rise to consistency conditions. As Geroch and Horowitz (1979) note, if we crop off enough of each end of the cylinder of Fig. X.2(b), any local solution of the source free Maxwell equations can be made global; intuitively, the electromagnetic disturbances, which propagate along null directions, go off the 'edge' of space-time before they have a chance to wrap all the way round the cylinder. But this model is both artificial and doubly undesirable, being not only acausal but singular as well (see Sec. 10 below).

But even if we grant that the argument does force the conclusion that the space-time structure of the actual world is causally nice, the doctrine of Laplacian determinism is not necessarily saved for the actual world. Again we need to remind ourselves that the doctrine quantifies over all physically possible worlds. So if there are non-actual but physically possible worlds with ugly causal structures, the actual world can fail to be Laplacian deterministic even though it is causally beautiful; we might inhabit a nice world like that of Fig. X.2(a) but have determinism wrecked by the possibility of the ugly world of Fig. X.2(b).

What then is the argument for requiring that all physically possible cosmological models have nice causal structures? The best known type of argument turns on the existence of paradoxes which are supposed to spring from acausal space-time structure; e.g., if there could be closed future directed timelike curves, an observer could travel into his own past and shoot himself-at-an-earlier-time thus preventing himself from living long enough to shoot himself. But such paradoxes do not demonstrate closed timelike curves are either logically self-contradictory or physically impossible. In fact, the paradoxes are simply a dramatic way of bringing out the point made above that acausal space-time features usually entail the existence of consistency conditions. And, although we may have some evidence against the existence of such conditions for the observationally accessible part of the actual world, I have yet to see a convincing argument against their realization in any physically possible world.

To appreciate the threat of acausality within the class of general relativistic cosmological models, two points need to be recognized in tandem. First, for some given gravitational source problems, the only known or the provably unique solution to Einstein's field equations exhibits acausal features (see Tipler (1974)). Second, these acausal features can be intrinsic to the model in the sense that they do not result from the kind of trickery used in Fig. X.2(b). More precisely, the acausal features are intrinsic if they cannot be removed by passing to a covering space-time. The closed causal loops of the space-time of Fig. X.2(b) are 'unwound' in passing to the space-time of Fig. X.2(a). By contrast, since Gödel space-time is simply connected, it is its own universal covering and so no unwinding is possible without doing damage to the local structure of space-time.

6. CAUCHY SURFACES

In order that the discussion not bog down on questions about acausality, let us agree to consider only world models whose causal structure is suitably well-behaved. In particular, it will be useful to assume at this juncture that there are global time slices which divide the space-time into three disjoint pieces: the slice S itself; the future $F(S)$ of S, consisting of all the points on the future side of S; and the past $P(S)$ of S, consisting of all the points on the past side of S. Then, as in the case of Minkowski space-time, we can say that $S \subset M$ is a *future* (respectively, *past*) *Cauchy surface* for M, g just in case $F(S) \subset D^+(S)$ (respectively, $P(S) \subset D^-(S)$), and that S is *Cauchy* simpliciter just in case it is both past and future Cauchy.

Three ways in which the Cauchy property can fail are illustrated in Fig. X.4. In Fig. X.4(a) the causal curve τ wraps endlessly around the universe without ever meeting S_a so that none of the points on τ belong to $D(S_a)$. But we have already agreed to ignore the acausal behavior involved in such an example, so let us turn to the other reasons for the non-existence of a Cauchy surface. Fig. X.4(b) is a schematic rendering of a feature of the Reissner-Nordstrøm cosmological model, representing the exterior gravitational field of an electrically charged, spherically symmetric body without angular momentum or dipole moment. S_b fails to be a Cauchy surface because the causal curve ρ does not register on S_b but emerges unpredictably from the singularity. Fig. X.4(c) illustrates the behavior of the null cones in the universal covering of anti-de Sitter space-time. The causal curve ω comes from spatial infinity without announcing itself on S_c, and in this way apes the behavior of the Newtonian invaders from infinity (recall Ch. III).

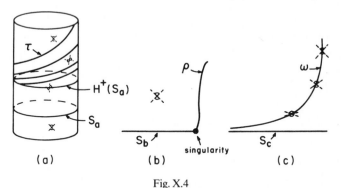

Fig. X.4

The anti-de Sitter cosmology can be regarded either as a solution to Einstein's field equations for empty space for $\Lambda = \frac{1}{4}R$, R being the constant (negative) curvature scalar, or else as a solution with $\Lambda = 0$ for a perfect fluid with negative density $R/32\pi$. Under the latter interpretation the model violates the energy condition imposed in Sec. 1. Under the former interpretation the model can be ruled out by the requirement that $\Lambda = 0$, a requirement for which there is both experimental and theoretical justification. However, the plane wave solutions of Penrose (1965) provide other examples of empty space solutions which are singularity free (at least in the sense of being geodesically complete), which do not contain closed causal curves, and which do not admit Cauchy surfaces. Of course, a pure plane wave solution is an idealization, and so one can wonder whether the relevant features carry over to more realistic solutions. And more generally, one can wonder whether there are any physically realistic solutions to Einstein field equations which are free of singularities and causal anomalies and yet do not possess Cauchy surfaces. If the determinist could establish a negative answer, his worries would be reduced to worries about singularities, a matter that will receive our attention below in Sec. 10.

If the space-time M, g does contain a Cauchy surface S, then M must be diffeomorphically $S \times \mathbb{R}$; in fact, there must be a diffeomorphism $d: M \rightarrow S \times \mathbb{R}$ such that $d^{-1}(S \times \{\lambda\})$, $\lambda \in \mathbb{R}$, are all Cauchy surfaces, and the map $t: M \rightarrow \mathbb{R}$ such that $t(d^{-1}(S \times \{\lambda\})) = \lambda$ is a global time function so that the space-time is stably causal. But if no Cauchy surface is present, the topology of space can change in the dramatic fashion illustrated in the trousers model of Fig. X.5. Let us say

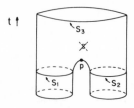

Fig. X.5

that M, g has an *upright trousers structure* iff M, g admits a global time function, and for every such function t there are constants c_1 and c_2 such that for all $\lambda < c_1$, the $t = \lambda$ time slices consist of two (or more) topologically disconnected components, whereas for all $\lambda > c_2$, the $t =$ time slices are connected.[4] Barring soothsaying and the like, the inhabitants of one 'leg' have no way of foreseeing that 'another world' will merge with theirs. Such a structure was once postulated for the actual universe in order to explain the appearance of new visible matter in the heavens (see Kundt (1967)).

7. THE CAUCHY INITIAL VALUE PROBLEM

Let us now focus on the class of general relativistic worlds which do possess Cauchy surfaces and see how determinism fares for these worlds. For simplicity, consider the initial value problem for the source free ($T = 0$) gravitational field. Here the initial data specify the state of the space-time itself, and since the space-time is not given *a priori*, the specification must come in a self-contained intrinsic form. A potential value set consists of a triple S, h, k, where S is a three manifold, h is a positive definite Riemann metric on S, and k is a symmetric tensor of type $(0, 2)$ defined on all of S. Intuitively, h, the 'first fundamental form' of S, describes the internal spatial geometry of S, while k, the 'second fundamental form,' describes how S is to be imbedded as a spacelike hypersurface of some space-time. Einstein's field equations (X.1)—(X.2) pose constraint conditions on the potential initial data. If these conditions are filled, S, h, k is called an admissible initial data set. One now wants to find a space-time M, g such that Einstein's field equations are satisfied (with $T = 0$), S is a Cauchy surface of M, g, and the first and second fundamental forms of S considered as a submanifold of M are h and k respectively. More precisely, a *Cauchy development* of the admissible initial data S, h, k is a triple M, g, θ where M, g is a space-time satisfying the field equations (with $T = 0$) and $\theta: S \rightarrow M$ is a diffeomorphism such that $\theta(S)$ is a Cauchy surface of M, g and the first and second fundamental forms of $\theta(S)$ are h and k. The principal result of the source free Cauchy initial value problem is this: for any given admissible initial data set there is a unique, up to a diffeomorphism, maximal Cauchy development (see Hawking and Ellis (1973)). The maximality of M, g means that it is an extension of any other such development. If M_1, g_1, θ_1 and M_2, g_2, θ_2 are both developments of S, h, k, then the second is said to be an extension of the first if there is a diffeomorphism $\Psi: M_1 \rightarrow M_2$ which isomorphically imbeds the first space-time into the second and which leaves S fixed (i.e., $\theta_1^{-1}\Psi\theta_2 = $ identity). The empty space initial value problem exhibits stability in that in an appropriate topology the solutions depend continuously on the initial data. Extending these results to include sources involves details of the dynamics of the sources and will not be discussed here (see Hawking and Ellis (1973) and Wald (1984)).

There have been attempts to put the Laplacian initial value problem for Einstein's field equations in a more classical form. From previous chapters we are accustomed to having initial data consisting of instantaneous values of basic quantities and their time derivatives, both of which are freely specifiable at the initial instant t_0. Thus, it is natural to try to take the gravitational initial data (for matter-free space) to consist of the inner metric $h(t_0)$ of the initial data hypersurface $S(t_0)$ and the time derivative $\dot{h}(t_0) = \mathrm{d}h/\mathrm{d}t|_{t_0}$ of h. With h and $\dot{h}$ specified, the constraint equations entailed by Einstein's field equations can be regarded as equations for the 'lapse' N_0 and the 'shift vector' N_α, ($\alpha = 1, 2, 3$) which together specify how a nearby hypersurface $S(t_0 + \mathrm{d}t)$ is to be constructed from $S(t_0)$. Wheeler (1964) conjectured that N_0 and N_α are uniquely determined by the constraint equations, thereby fixing how $S(t_0)$ is to be imbedded in space-time. Unfortunately, this 'thin sandwich' conjecture proves to be false, as both existence and uniqueness can fail, so that only under very special conditions can the Laplacian initial data be divided into a part freely specifiable (h and $\dot{h}$) and a part (N_0 and N_α) uniquely determined by the constraint equations (see Belasco and Ohanian (1969) and Christodoulou and Francaviglia (1979)).

8. THE SIGNIFICANCE OF THE CAUCHY INITIAL VALUE PROBLEM

Several features of the initial value problem call for further comment. The uniqueness result holds only for maximal developments, because for any given development of S, we can produce another by deleting a closed set of points from the manifold in such a way that S remains a Cauchy surface. A trivial illustration is given in Fig. X.6. S is a Cauchy surface for the truncated space-time that results from deleting all the points on or above the time slice $t =$ high noon on April 1, 1988. Such truncated models offend deepseated metaphysical intuition: there is no sufficient reason why Nature (or God if you prefer) should stop building at 1988. This intuition can be codified in the requirement that any acceptable space-time model must be inextendible, i.e., not isometrically imbeddible as a proper subset of another space-time. But whatever the validity of this intuition as regards global inextendibility,

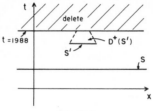

Fig. X.6

we will see below that our intuitions become strained when we turn to the stronger inextendibility conditions needed to support localized versions of Laplacian determinism.

The second aspect of the uniqueness result that needs emphasis is the 'up to a diffeomorphism' proviso. I claimed that the major revolutions of 20th century physics were not caused by self-conscious reflections on the problems and prospects of determinism. That claim is correct in the sense that these revolutions were neither initiated nor brought to fruition by such reflections. But reflections on causality and determinism did delay Einstein's discovery of the gravitational field equations. In 1913 he tried but failed to find acceptable generally covariant equations. What he could not do, Einstein became convinced, could not be done at all. The culprit, he decided, was general covariance, and to show why he concocted an argument to the effect that general covariance is incompatible with the determination of the gravitational field from the distribution of matter-energy. The result was a wild goose chase, lasting over two years, after non-covariant field equations (see Norton (1984)).

Einstein's instincts were partly right and partly wrong. General relativity does mark a break with the classical formulation of the doctrine of determinism, but the break is not due to general covariance and it does not signal a demise of determinism. The laws of classical and special relativistic physics can be written in generally covariant form, but this mode of presentation makes for no significant difference for determinism. It is rather the absoluteness of the space-time, the fact that the space-time structure remains similar and immovable, that makes possible the description of the classical and special relativistic initial value problems against a fixed scaffolding. Whether the scaffolding allows a freedom to perform diffeomorphisms which is harmful to Laplacian determinism is a matter that has to be examined on a case by

case basis. In Minkowski space-time the freedom is effectively
squelched since any global isometry which reduces to the identity on a
(Cauchy) initial value surface is the identity everywhere. By contrast, a
symmetry mapping of Leibnizian space-time (Ch. III.2) can be chosen
to reduce to the identity on a neighborhood of a plane of simultaneity
but differ from the identity outside of the neighborhood. To rescue
Laplacian determinism either the structure of Leibnizian space-time has
to be beefed up by passing, for example, to Newtonian space-time, or
else the space-time models have to be interpreted along the lines
suggested by Leibniz in his correspondence with Clarke (Ch. III.3). In
the case of the general theory of relativity only the latter option is open
since the theory banishes absolute structure and the freedom to
perform diffeomorphisms is unlimited. As a result, if M, g solves
Einstein's source free field equations and possesses a Cauchy surface S,
then it is easy to generate another solution M, g' such that S is also a
Cauchy surface for M, g' and such that $g = g'$ on a neighborhood of S
but $g \neq g'$ otherwise. For let d be any diffeomorphism of M onto itself
which is the identity in some neighborhood of S but not otherwise and
choose $g' = d^*g$ (see Earman and Norton (1986) for details). Thus, to
save determinism the diffeomorphically related models must be
regarded as merely different modes of presentation of the same physical
reality. As indicated in Ch. III, I believe that an intrinsic description of
the underlying reality necessitates performing the trick of doing without
the manifold M as a point set while retaining the differentiable structure
and the structure of the g-field. This is perhaps what Einstein had in
mind when he wrote:

If we imagine the gravitational field, i.e. the functions g_{ik}, to be removed, there does not
remain a space [of the relativistic type], but absolutely *nothing*, and also no 'topological
space' . . . There is no such thing as empty space, i.e. a space without a field. Space-time
does not claim an existence of its own, but only as a structural quality of the field.
(1961, p. 155).

The combination of general relativity and a faith in Laplacian
determinism makes an implementation of the Leibniz-Einstein vision
imperative. Physics has accomplished what Leibniz' Principle of
Sufficient Reason could not.
    The third comment concerns the status of the requirement that S is a
Cauchy surface. Recall the remark from Ch. IV that the statement that
S is a Cauchy surface for M, g is a statement about the entirety of M, g.
That caused no embarrassment in the setting of special relativity theory

where M, g is always Minkowskian and where Cauchy surfaces exist and can be recognized as such by local characteristics. But in the context of general relativity the hypothesis that S is Cauchy is potentially pernicious, for it cannot be tested by measurements made only on or near S and it rules out in advance various possibilities allowed by Einstein's field equations. To assume that S is Cauchy is thus to make a substantive assumption to the effect that the space-time structure is not such as to harbor unsettling surprises; and that is an embarrassment since it assumes what determinism was supposed to guarantee. The embarrassment would be overcome if the existence of a Cauchy surface were made an entrance requirement for the class of physically possible world models. Arguments in support of that requirement will be discussed in Sec. 11 below.

9. LAPLACIAN DETERMINISM IN THE MEDIUM AND THE SMALL

Most of the challenges to Laplacian determinism discussed so far arise from large scale features of space-time structure and, thus, are relevant only to determinism as it concerns the evolution of the entire universe. This naturally raises the question of whether a more modest form of determinism as applied to the local scene can hold even if determinism breaks down in the large. In particular, if the spacelike hypersurface S is not Cauchy for M, g then the state on S cannot determine the state throughout M, g; but it would be nice to be able to say that the state on S, whether a partial or full time slice, determines the state throughout its domain of dependence $D(S)$, which may be a proper subset of M. This will be true only if we are dealing with a space-time M, g where for each spacelike S the maximal Cauchy development of S is attained within M, g. More precisely, the condition is that M, g be *hole free* (Geroch (1977)), which obtains just in case for any spacelike S (assumed to be achronal) there do not exist a space-time M', g' and an isometric imbedding $\Psi: D(S) \rightarrow M'$ which makes $\Psi(D(S))$ a proper subset of $D(\Psi(S))$.

The truncated space-time of Fig. X.6, Minkowski space-time with a compact ball removed, and other similar examples are obvious illustrations of non-hole free space-times. But these examples are already ruled out by our previous agreement to consider only maximal or inextendible space-times, so it might seem that we have already made general relativistic worlds safe for local determinism. Alas, there are two rubs here. The first is a potential conflict between the demands of maximality and hole freeness. The maximality demand can always be met since any space-time can be imbedded in another space-time which itself is not properly extendible. But not every hole free space-time has a hole free maximal extention (see Clarke (1976)). Perhaps there are plausible regularity conditions on initial data which will rule out such latent holes, but I know of no specific results on this point. The second rub is that maximality does not guarantee hole freeness. Start with Minkowski space-time and remove the two-plane of points $x = 0$, $t = 0$ where (x, y, z, t) is an inertial coordinate system and then take the universal covering space-time. The end product is inextendible but not hole free (Clarke (1976)). Strengthening the inextendibility demand by imposing local as well as global inextendibility will banish this counter-example. Following Hawking and Ellis (1973), let us say that M, g is locally extendible just in case there is an open $U \subset M$ with non-compact closure in M and a space-time M', g' and an isometric imbedding $\Psi: U \to M'$ such that $\Psi(U)$ has compact closure in M'. Local inextendibility does guarantee hole freeness, but unfortunately the price for this guarantee is too high since, for example, standard Minkowski space-time is locally extendible, as shown by the construction of Beem (1980) illustrated in Fig. X.7. A more stringent form of

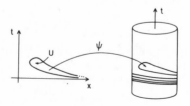

Fig. X.7

local inextendibility, called local b-inextendibility (see Sec. 10 below) escapes this embarrassment while still guaranteeing hole freeness. However, it is not clear that this and other local maximality requirements are acceptable *a priori* demands on space-time models. And I know of no place where research workers in general relativity use such requirements to justify hole freeness; indeed, the typical procedure is a direct appeal to hole freeness on the grounds that we should not tolerate a breakdown in determinism due to a capricious appearance of uncaused singularities (see Ellis and King (1974), Clarke (1976), and Ellis and Schmidt (1977)). As always, fiat stands ready to establish determinism where honest toil does not suffice.

All of these worries can be avoided by retreating to a small enough level. For any cosmological model M, g, T and any point $x \in M$, we can always find a small enough neighborhood U of x such that $U, g|_U$ possesses a Cauchy surface and $U, g|_U, T|_U$ is a maximal, and therefore, unique Cauchy development of the initial data on S. The usefulness of this triumph of determinism in the small depends upon how small small is. The resultant sense of determinism will be epistemologically useless if the existential clause is filled only by regions so minute as to be irrelevant to typical prediction problems. And in any case, we may have no way of knowing in advance how large or small the region is. Ontologically, determinism in the small does not sustain James' vision of a world in which the womb of the future contains no ambiguities. The myriad of miniature subworlds within which James' vision is fulfilled may not join together into a Jamesian absolute unity in which there is no equivocation or shadow of turning.

10. SINGULARITIES

The topic of singularities is at once one of the most exciting and vexing in general relativistic physics. A good part of the vexation comes from the difficulty of capturing intuitions about singularities in a precise and tractable definition. Intuitively, a singular point in space-time is a place where the space-time metric becomes singular, e.g., undefined or non-differentiable. But such points can be excised from the manifold, leaving a space-time where the metric is everywhere regular. In effect, our definition of a relativistic space-time M, g assumes that such excisions have been made since g is assumed to be defined on all of M.

It thus appears that the key question in determining whether a given space-time M, g is singular is whether any points have been omitted. Extendibility signals that regular points have been omitted. Since any space-time can be imbedded in an inextendible space-time, we may therefore concentrate on maximal space-times, and our task reduces to that of devising a criterion for detecting the omission of non-regular points in these space-times. The presence in a maximal space-time of geodesics which cannot be extended to arbitrarily large values of an affine parameter is generally regarded as sufficient to signal a true singularity. But it is apparently not necessary since there are geodesically complete space-times which contain inextendible timelike curves of bounded acceleration and finite proper length (Geroch (1968)). The unfortunate observer whose rocket ship traces out such a curve would surely have just as much right to regard his finite life span as due to a singularity as would an observer who free falls along an incomplete timelike geodesic. A still stronger condition of completeness, called *b-completeness*, requires that every C^1 curve be extendible to an arbitrarily large value of a generalized affine parameter.[5] But one can wonder whether *b*-incompleteness in maximal space-times is a sure-fire sign of a genuine physical singularity. The answer depends, of course, on what we count as a 'physical' singularity. But we can make the problem more precise by asking: Might not the curvature of space-time remain well behaved along a *b*-incomplete curve? Could it not be that the missing points are quasi-regular in that they can be recovered by local extensions along the *b*-incomplete curves? In the space-time M, g let $\sigma(\lambda)$, $\lambda \in [0, r)$, be a *b*-incomplete curve which cannot be extended beyond $\lambda = r$. Then M, g is said to be *locally extendible along* σ just in case there is an open neighborhood $U \subset M$ of σ, a space-time M', g' and an isometric imbedding $\Psi: U \to M'$ such that $\Psi(\sigma(\lambda))$ is continuously extendible beyond $\lambda = r$. Clarke (1973) shows that such an extension is possible just in case the components of the Riemann curvature tensor taken in a frame parallel propagated along σ approach a limit as $\lambda \to r^-$. Thus, the singularities involved are labeled *quasi-regular*.

The non-quasi-regular or *curvature singularities* can be classified in various ways, depending upon whether or not a curvature scalar fails to approach a limit, whether the limit fails because of blowup or oscillatory behavior, etc. Details of the classification and theorems on the existence of singularities in general relativistic models are reviewed in articles by Ellis and Schmidt (1977) and Tipler, Ellis, and Clarke (1980). Suffice it to say here that a series of results, initiated by Hawking and Penrose, establishes that the existence of space-time singularities in the sense of geodesic incompleteness is a highly generic feature of the cosmological models of general relativity. Results on the existence and nature of curvature singularities are more piecemeal.

Singularities can entail a breakdown in determinism, as indicated in Fig. X.4(b). But singularities can peacefully coexist with local and global determinism; indeed, a space-time can be singular and still admit

Cauchy surfaces, the Kruskal maximal extension of the Schwarzschild solution (Fig. X.8) and various 'big bang' cosmologies of the Robertson-Walker type being relevant examples. The cosmic censorship hypothesis, discussed in the following section, can be read as an attempt to show that in reasonable models of cosmology and of gravitational collapse such peaceful coexistence is the norm. But before turning to that topic, something must be said about an issue that has been skirted.

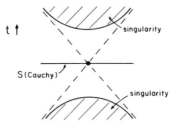

Fig. X.8

The discussion thus far has been deliberately ambivalent over the issue of whether singularities are legitimate objects of physics. Our definition of a general relativistic space-time implies that they are not fully legitimate objects since they are not part of space-time. To this we could add the viewpoint that they are wholly illegitimate objects of scientific enquiry; for singularities imply a breakdown in the picture of space-time as a Lorentzian manifold, and since all of the currently known laws of physics presuppose this picture, the enterprise of physics as it is currently practiced is condemned to silence as to what transpires at a singularity. Some would go further and add that entities that are not the subjects of scientific enquiry cannot occur in nature. The Hawking-Penrose theorems, showing that singularities are a pervasive feature of general relativistic models, then become part of a transcendental proof of the falsity of the general theory of relativity. The most prevalent conjecture as to what goes wrong is that the singularities (in the sense of incompleteness) are typically associated with regions of unbounded curvature and that in these regions quantum effects, which are neglected in the classical theory, become dominant. However, it is far from clear that quantized gravitation will avoid space-time singularities (see Wheeler (1977)). And even if singularities are avoided, it is likely that the amalgamation of quantum theory and the general theory of relativity will result in a radically new conception of space-time; whether or not this new conception will permit the application of Laplacian determinism in a form that is recognizably akin to its classical form is now a matter of pure speculation.

The alternative viewpoint takes general relativity at face value and accepts the conclusion that, most probably, space-time singularities occur in the actual universe, either at the initial big bang or at the end of gravitational collapse. Of course, these singularities do not literally occur as events in space-time. But the next best thing would be to speak of them as occurring at ideal points attached as boundary points of the space-time manifold. Unfortunately, all of the attempts to construct such boundaries have unsatisfactory features (see Tipler *et al.* (1980)). It remains to be seen whether the deficiencies can be remedied by technical modifications of the constructions or whether the deficiencies indicate that space-time singularities are too intractable to admit a satisfactory characterization as limit points of ordinary space-time. In either case nothing more can be said about the happening at these ideal points, suggesting that functionally nothing much rides on whether we say that singularities 'really' occur in nature.

Even when they do not clash with the existence of a Cauchy surface, singularities are an ugly stain on the success of determinism in general relativity. Focus on the subclass of models with Cauchy surfaces. Then by our definition of determinism and the results of the gravitational initial value problem, Laplacian determinism holds. But for models with singularities the victory of determinism has a Pyrrhic flavor, for at best the prediction of singularities is a prediction of the breakdown of the laws of the theory. That breakdown is not countenanced as a breakdown in determinism since the 'places' where the singularities occur are not countenanced as part of the arena where determinism wins or loses. The ever more clever means by which determinism avoids falsification are as impressive as its straightforward successes.

11. COSMIC CENSORSHIP

In 1978 Roger Penrose wrote that "possibly the most important unsolved problem of classical general relativity theory" is the cosmic censorship question: Does GTR enforce cosmic censorship, forbidding singularities that develop in physically realistic cases of gravitational collapse to appear 'naked' to the world? Today the problem remains

unresolved, partly because of the difficulty in defining naked singularities and partly because of the difficulty in separating 'realistic' from
'unrealistic' behavior. There is a direct connection between cosmic
censorship and determinism; indeed, the strong cosmic censorship
hypothesis (CSH) is sometimes formulated in terms of the existence of
a global Cauchy surface while the weak CSH is sometimes formulated
in terms of a Cauchy surface for the exterior of a black hole.

To see what is at stake in the CSH it is helpful to run through a
series of examples, starting with Schwarzschild-Kruskal space-time (Fig.
X.8), the Robertson-Walker 'big-bang' models (indicated in schematic
from in Fig. X.9(a)), and the various gravitational collapse models
illustrated in Figs. X.9(b)—(e). In the first two cases, we can 'see' (with

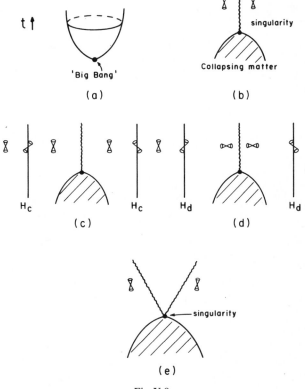

Fig. X.9

the help of microwave antennae) the initial singularity, but we do not regard this as a violation of the CSH, strong or weak, since the singularities were there at the beginning of time. By contrast, the singularity of collapse that develops in case (b) is as naked as a babe, violating any reasonable form of the CSH. In cases (c) and (d) nature has exerted some modesty in hiding the singularities behind the event horizons[6] labeled with Hs. These horizons serve as one-way causal membranes, allowing influences to go in but none to come out; in short, both cases contain black holes. The difference between (c) and (d) is that in the former case an observer who enters the black hole can see the singularity while his unfortunate counterpart in case (d) is snuffed by the singularity before he gets a chance to view it. Summarizing, we would like to be able to say that the Schwarzschild-Kruskal space-time, the Robertson-Walker 'big bang' models, and the collapse model (d) all illustrate the strong CSH; while model (c) satisfies the weak but not the strong CSH; and model (b) violates the weak CSH.

A means of capturing these intuitions is supplied by a construction of Geroch and Horowitz (1979). Define the *naked viewing set* N of a space-time M, g to be the collection of all points $p \in M$ such that there is a $q \in M$ which precedes p and whose own causal past entirely contains a timelike or null curve γ which is maximally extended in the future but which has no future end point. The existence of such a γ indicates that some form of singular behavior develops; what kind of singularity depends upon what further conditions we impose on γ, e.g., we could require that γ is future b-incomplete, that the space-time is locally b-inextendible along γ, that a curvature scalar blows up along γ, etc. The rest of the definition of N guarantees that this singular behavior is not veiled from observation by an external observer. For Schwarzschild-Kruskal space-time, the big-bang models, and example (d), $N = \emptyset$. In cases (b) and (c) $N \neq \emptyset$, but N is empty for the complement of the black hole in (c).

Thus, we are encouraged to try the following formulation of cosmic censorship. M, g satisfies the strong CSH just in case $N = \emptyset$; and it satisfies the weak CSH just in case $N = \emptyset$ for $E, g|_E$ where $E \subset M$ is the exterior of a black hole. It is easy to see that the existence of a global Cauchy surface implies that the strong CSH holds; and likewise the existence of a Cauchy surface for each connected component of E, $g|_E$ implies that the weak CSH holds. And if N is defined without any further requirement on γ other than that it is maximally future extended but trapped in the causal past of q, then the existence of a Cauchy surface is necessary as well as sufficient for the validity of the strong CSH (see Geroch and Horowitz (1979)).

Penrose (1978) has questioned whether the proffered form of the strong CSH is strong enough. In example (e) collapsing matter produces a singularity that emits an infinite pulse of radiation that destroys the universe as it goes. Isn't this a case of a cataclysmic naked singularity *par excellence* even though $N = \emptyset$? (e) is not literally a counterexample since if the pulse of radiation does produce a real curvature singularity that cuts off future development, then it is not part of the space-time, and the singularity cannot literally be seen though it can be 'seen to be coming.'

More worrisome is whether our formulation of cosmic censorship is too strong. In Taub-NUT space-time, some of whose causal features are rendered in Fig. X.4(a), $N \neq \emptyset$, but it is not clear that the kind of singular behavior exhibited counts as a violation of cosmic censorship. But this need not cause embarrassment if we assume, as we have already, that acausal space-times have been excluded from the realm of the physically possible. Another potential embarrassment comes from the anti-de Sitter space-time (Fig. X.4(c)) which is stably causal and singularity free and yet has a non-empty naked viewing set on the most liberal way of defining N. This suggests that we should stick to the more restrictive forms of definition on which γ is required at the least to be future b-incomplete. If this suggestion is taken to heart the existence of a Cauchy surface is not a necessary condition for the validity of the CSH.

However, the existence of a Cauchy surface is (*pace* Penrose's qualms) sufficient for the validity of the CSH, and so it is not unexpected that the proponents of censorship have offered to convince us that a Cauchy surface should exist. The *future Cauchy horizon* $H^+(S)$ of the spacelike slice S is defined as the future boundary of $D^+(S)$ (i.e., the closure of $D^+(S)$ minus the chronological past of $D^+(S)$). Intuitively, $H^+(S)$ separates the part of the future determined by the state on S from the part not so determined. Of course, S is future Cauchy just in case $H^+(S) = \emptyset$. Now consider a space-time which does not contain a Cauchy surface. Choose an appropriate slice S, for which by assumption $H^+(S) \neq \emptyset$, and try to show that a small change in the initial data on S destroys $H^+(S)$. In the case of the Taub-NUT space-time (Fig. X.4(a)) a dust cloud introduced on S

accumulates near $H^+(S)$, creating a singularity that destroys $H^+(S)$ and cuts the lower part of the space-time off from the upper part. Analogous instabilities are also found in the Reissner-Nordstrøm space-time and the anti-de Sitter model. However, all of these are examples of homogeneous cosmologies and not much is known about the inhomogeneous cases. My own guess is that perturbations in inhomogeneous models won't in general destroy Cauchy horizons without creating new ones.[7]

Even if it could be shown that Cauchy horizons in cosmology are unstable, a leap is needed to get to the conclusion that models of cosmology without Cauchy surfaces are beyond the pale of the physically possible, or else that they are ignorable because they are of measure zero or because of some other reason. It is instructive to compare the situation here with that in ergodic theory (see Ch. IX). For hard spheres in a rectangular box with perfectly reflecting walls, there are initial states that do not lead to mixing or even quasi-ergodic behavior; imagine, for instance, that the particles are initially arranged so that they follow non-intersecting parallel trajectories that are perpendicular to the walls of the box. Not only are such initial conditions exceptional (measure zero), but they are also unstable under external perturbations. No attempt to isolate the box will be proof against such perturbations; there is no shield against gravitational attraction, and the attraction of even the distant stars may be enough to wreck the coherence of the initial trajectories and reinstate mixing. (Maybe there is something to astrology after all.) This double improbability of non-mixing behavior is not equivalent to physical impossibility, but it comes close. In cosmology there is no corresponding double improbability since the universe is all there is and so there can be no external perturbation.

Turning now from cosmology to gravitational collapse, which after all was the initial focus of the CSH, the evidence is more ambiguous. Yodzis et al. (1974) have shown that naked singularities violating the weak CSH can emerge from a spherically symmetric collapse of a fluid and that the result is stable under small but still spherically symmetric changes in the initial data and under small changes in the equation of state. Nor is the result attributable to the spherical symmetry, as is shown by Szerkeres' (1975) non-spherically symmetric model. However, these cases have been dismissed as spurious counterexamples to the CSH on the grounds that they are 'unphysical' with the singularities being merely artifacts of the continuum approximation (see Hawking (1979)). Details of this debate about what equations of state are needed to qualify a model of gravitational collapse as being physically realistic are too technical for us to follow here, but we should not be surprised at the existence of the debate; for every time there is an apparent breakdown of determinism an attack is made on the reasonableness of the determinism breaking examples.

12. PREDICTION

Special relativity (sans tachyons) made the world safe for determinism by screening off the invaders from spatial infinity. But the very space-time structure which screens off the invaders also screens off observers from causal contact with past events in such a way as to make prediction impossible. We have seen that the message general relativity holds for determinism is mixed. As we will now see, the message for prediction is equally mixed.

The domain of prediction for each point in Minkowski space-time is empty. The same is true of many of the cosmological models of general relativity, e.g., the Robertson-Walker models which have been used to describe a universe beginning or ending with a 'big bang.' But in some cases prediction is possible on a grand scale. In the best of all models the space-time M, g is such that for each $x \in M$, there is a Cauchy surface $S \subset C^-(x)$. All observers everywhere are always in a position to obtain enough information to predict the future of the entire universe. These best of all predictable worlds must be spatially finite, for it can be shown for a Cauchy surface S to be contained in the causal past of a point entails that S must be compact. The converse is not true; not every model with a compact Cauchy surface is a best predictable world. The de Sitter cosmological model is a case in point. The behavior of the null cones of de Sitter space-time is illustrated schematically in Fig. X.10. Note that S is a Cauchy surface, but S is not contained in $C^-(x)$ for any point x. In fact $DP(x) = \emptyset$.

It is natural to wonder whether there are cases where some prediction is everywhere and always possible, but where there is neither a Cauchy surface nor the possibility of global prediction. The answer depends upon how extensive the domain of prediction is required to be. If it is required only that it be non-empty, then, as illustrated in Fig. X.11, there is a space-time M, g where for every $x \in M$, $DP(x) \neq \emptyset$,

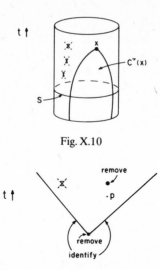

Fig. X.10

Fig. X.11

but M, g has no Cauchy surface. This model, however, has the curious feature that the domain of prediction of the point p contains all events having a spacelike relation to it, but many of the events in the causal future $C^+(p)$ are not in p's domain of prediction. Since the events an observer is most likely to want to predict are those in which he can participate or influence, it is natural to wonder about the requirement that for every $x \in M$, $C^+(x) \subset DP(x)$. Budic and Sachs (1976) show that under reasonable causality conditions this requirement entails that $C^-(x)$ contains a Cauchy surface, which by the above results must be compact.

Much more could be said about prediction in general relativistic worlds, but the examples given already are enough to motivate the 'meta-theorem' that for almost any hypothesis one can dream up about predicton, there is a cosmological model to illustrate it.

13. GEOMETRODYNAMICS

The successful theories of physics, from Newton's time to the present, have used or tolerated various dualisms, such as particles and fields. But until this century, the most persistent and seemingly unbridgeable dichotomy has been that which separates the G's (the geometry of space-time) from the P's (the physical contents) of the world models $\langle M, \{G_\alpha\}, \{P_\beta\}\rangle$. Einstein's general theory moved in the direction of absorbing the P's into the G's; in particular, this theory banishes gravitational force and views the gravitational field as an aspect of the space-time geometry rather than as a separate physical field. However, on the orthodox version of general relativity, space-time is still colonized by various matter-energy fields as described by the stress energy-tensor T. Geometrodynamics is a program whose avowed goal is to end this colonialism either by doing without the P's or else reducing them to metrical and/or topological aspects of space-time ('curved empty space-time is all there is').

My concern here is not with the general desirability of plausibility of such a program but rather with the role of determinism in any acceptable realization of it. As a case in point, the Misner-Wheeler 'already unified' theory of gravitation and electromagnetism appeared at first to carry the program a step forward by showing how, within the bounds of Einstein's general theory, the electromagnetic field could be identified with features of the space-time geometry. However, the initial value problem in this theory argues against such an identification, at least if determinism is to be maintained. For purely geometric measurements made on or near a spacelike hypersurface are not sufficient to determine a unique development of the electromagnetic field in the Misner-Wheeler scheme. This failure of determinism was taken by proponents and critics of the theory alike as an indication that the sought after unification had not been achieved and that the electromagnetic field remained a colonizing entity (see Wheeler (1962, 1968)).

14. THE CHARACTERISTIC INITIAL VALUE PROBLEM

In Ch. II we noted the possibility of non-Laplacian varieties of determinism, but subsequently our focus has been drawn almost exclusively to the Laplacian form. By way of justification we can point to the facts that when philosophers discuss the meaning, the truth, or the implications of determinism it is almost always Laplace's version, or a close cousin, they have in mind and that most of the standard mathematical analysis of the existence and uniqueness of solutions to the equations of physics deals with the case of instantaneous initial data. It would be remiss, however, to close the discussion of determinism in relativistic physics without at least mentioning that in this context a characteristic initial value problem, with data given on a null surface, is in some ways more natural and attractive than a Laplacian type initial value problem.

Null surfaces are more natural and important, both ontologically and epistemologically, than spacelike surfaces; for electromagnetism and gravity propagate along null directions, and much of our information about the far reaches of the universe comes to us on a series of null surfaces. The characteristic initial value problem for Einstein's field equations has been studied by Sachs (1962), Penrose (1967), and Müller zum Hagen and Seifert (1979). They find that a bonus of this approach is that, in contrast to the Laplacian initial data which must meet constraint conditions (see Sec. 7 above), the characteristic initial data needed to determine a unique solution can be freely specified. At the same time, their results raise a puzzle about how to count degrees of freedom of the gravitational field, for four independent pieces of Laplacian data (four real numbers at each point) are needed for a well posed problem, whereas only two pieces are needed in the characteristic problem. This difference is not due to differences in domain of dependence, since, for example, the past null cone and the spacelike S_3 in Fig. IV.2(c) have exactly the same domains of dependence (the interior of the past cone).

In the future direction of time there is a sharp break between predictability and determinism via null hypersurface data; for to determine the state, say, in the interior of the future lobe $\mathcal{N}_p^+$ of the null cone[8] through a point p, null data must be given for all of $\mathcal{N}_p^+$ (or some other null surface extending into the future in such a way as to capture $\mathcal{N}_p^+$ in its domain of dependence), and such data are inaccessible to any observer existing at the present moment. But by now the wedge between predictability and determinism has been driven in far enough that this break can hardly be counted as a reason for dismissing the characteristic initial value problem as a mathematical artifice. The feeling of artificiality may also derive from the metaphysical intuition that the future is extruded from the present so that only the Laplacian form of determinism properly reflects the true ontological unfolding of events. It is, however, difficult to point to anything in the laws of physics to support this intuition, and in any case it is hard to square the intuition with the structure of relativistic space-times which make no provision for an invariant notion of the instantaneous present.

15. CONCLUSION

While determinism in the small is a certainty in general relativistic worlds, determinism in the medium and the large remains an open question. Additional observational and theoretical results could help to resolve some of the remaining uncertainty; but the ultimate fate of large scale determinism turns on some sticky interpretations problems about what counts as a reasonable space-time model, and these problems resist narrowly scientific solutions. If this is true for the general theory of relativity, then it is doubly true for the quantum theory, as we will now see.

NOTES

[1] Thus we disregard here the possibility of tachyons.

[2] For definitions and mathematical details, see Hawking and Ellis (1973). In some formulations of general relativity theory Λ is required to be 0.

[3] Should time in the space-time of Fig. X.3 be considered 'open' (because of the existence of a global time function) or 'closed' (because of the seemingly circular structure of the time dimension)? Draw the level surfaces of a global time function for this space-time.

[4] Fig. X.5 is only schematic, for there is trouble at the crotch point p in defining a Lorentz metric.

[5] Let $V^i(\lambda)$ be the components of the tangent vector $\dot{\sigma}(\lambda)$ and let (E_1, E_2, E_3, E_4) be a frame of linearly independent vectors parallel propagated along $\sigma(\lambda)$. Then the generalized affine parameter is defined by

$$s \equiv \int_0^\lambda \left[\sum_{i=1}^4 (g_{jk} V^j E_i^k)^2 \right]^{\frac{1}{2}} d\lambda$$

The curve $\sigma(\lambda)$ is finite in s iff it is finite in any other s' obtained by using another frame. If $\sigma(\lambda)$ is a geodesic, then s is an affine parameter.

[6] The (future) *event horizon* of an observer O (as represented by a timelike worldline) is the boundary of O's past. In Minkowski space-time, an observer who is unaccelerated and who lives forever (no future end point to O's worldline) has an empty event horizon — O's past light cone eventually sweeps out all of Minkowski space-time. However, there are in Minkowski space-time timelike curves which have no future end points and yet have non-empty evnt horizons. (Reader: Draw such a curve.) In an asymptotically flat space-time, *the* event horizon is defined as the boundary of the past of 'future null infinity' (roughly, the boundary of the events which can be 'seen' from near future infinity); if this boundary is non-empty there are said to be 'black holes' present (see Hawkings and Ellis (1973)).

[7] For some contrary evidence, see Moncrief (1982).

[8] Here $\mathcal{N}_p$ refers to the global null cone at p, generated by the null geodesics through p.

SUGGESTED READINGS FOR CHAPTER X

A good introducton to the issues discussed in this chapter is to be found in Geroch (1977) "Prediction in General Relativity." Comprehensive treatments of the relevant mathematics and physics are given in Hawking and Ellis (1973) *The Large Scale Structure of Space-Time* and Wald (1984) *General Relativity.* A number of useful review articles are to be found in the collections by Hawking and Israel (1979) *General Relativity*, and Held (1980) *General Relativity and Gravitation.*

CHAPTER XI

DETERMINISM IN QUANTUM PHYSICS

> ... I should not want to be forced into abandoning strict causality without defending it more strongly than I have so far. I find the idea quite intolerable that an electron exposed to radiation should choose *of its own free will*, not only its moment to jump off, but also its direction. In that case I would rather be a cobbler, or even an employee in a gaming-house, than a physicist.
>
> (A. Einstein to M. Born, April 29, 1924)

There is an ironic twist to the timing of Einstein's letter to Born: during the following two years Schrödinger and Heisenberg created the formalism of the 'new quantum theory' and in 1926 Born proposed his statistical interpretation of the formalism, work for which he received the Nobel Prize. Over the years the divergence between Einstein and Born in their attitudes towards the quantum theory continued to grow. In 1944 Einstein wrote to Born that "We have become Antipodean in our scientific expectations. You believe in the God who plays dice, and I in complete law and order in a world which objectively exists ..." (Born (1971), p. 149). By the late 1940s a tone of exasperation began to creep into Einstein's remarks to Born on the foundation of quantum mechanics. At one point Einstein wrote, with wholly uncharacteristic asperity, "I do not want to take part in any further discussion, such as you seem to envisage. I content myself with having expressed my opinion clearly." (Born (1971), p. 212). The intercession of Wolfgang Pauli was required to put the discussion back on track.

Einstein's uneasiness with the quantum theory was not simply a function of his conviction that at base the world is not a dice game, though that was a frequently voiced complaint. The theory grated against several ontological and methodological principles Einstein held to be fundamental, and the failure of his friend Born to appreciate the force of these principles and Born's tendency to try to reduce all objections to a failure of classical causality was to Einstein a source of annoyance. Pauli reported to Born that "Einstein does not consider the

199

concept of 'determinism' to be as fundamental as it is frequently held to
be (as he told me emphatically many times) ... Einstein's point of
departure is 'realistic' rather than 'deterministic' ..." (Born (1971), p.
221). Nevertheless in trying to understand the status of determinism in
quantum physics we will, inexorably, be brought face to face with most
of Einstein's worries.

Since my main concern has been with ontological determinism,
previous chapters on the implications of various parts of classical and
relativistic physics have begun with a sketch of the assumed ontological
world structure. It is impossible to begin this chapter in a similar
manner, for the nature of quantum ontology is the locus of the most
basic and controversial of unresolved foundations problems. All that
can be said initially is that ordinary non-relativistic quantum mechanics,
the main focus of this chapter, does nothing to change the classical
space-time assumed in Ch. III. How to characterize what goes on within
that framework will prove to be a key issue. We will have to grope our
way towards an answer.

1. QUANTUM MECHANICS AS MORE DETERMINISTIC THAN
CLASSICAL MECHANICS

Ernest Nagel has noted that "relative to its own form of state description quantum theory is deterministic in the same sense that classical
mechanics is deterministic with respect to the mechanical description of
state" (Nagel (1961), p. 306). There are important caveats to be
discussed below, but to the extent that Nagel is right he understates
his case: quantum mechanics is *more* deterministic than classical
mechanics.

The Schrödinger equation for a single particle moving in an external
potential $V(q)$ in one spatial dimension labeled by q reads

$$(XI.1) \quad i\hbar \frac{\partial \psi}{\partial t} = - \frac{\hbar^2}{2m} \frac{\partial^2 \psi}{\partial q^2} + V\psi$$

where $\psi(q, t)$ is a complex valued function variously called the 'state
function', the 'wave function', or just the 'psi function'. In the case of a
free particle (XI.1) reduces to

$$(XI.2) \quad i\hbar \frac{\partial \psi}{\partial t} = - \frac{\hbar^2}{2m} \frac{\partial^2 \psi}{\partial q^2}$$

which bears a superficial resemblance to the classical heat equation studied in Ch. III. There are, however, major differences. Unlike the heat equation, (XI.2) does not smooth out initial data, and while the heat equation is not time reversal invariant, (XI.2) is, at least if we adopt the convention that the state reversal operation (see Ch. VII) is given by $[\psi(q, t)]^R = \psi^*(q, t)$, where '*' denotes complex conjugation.

But a crucial point of contact is that both the heat equation and (XI.2) are of the parabolic type and thus allow infinitely fast propagation of disturbances. In the case of the heat equation this feature created problems for Laplacian determinism because of the possibility of disturbances coming in from spatial infinity. In the case of quantum mechanics $\psi(q, t)$ will be interpreted, following Born, as a probability amplitude, and this supplies the boundary condition at infinity sufficient to prove uniqueness: for probability to normalize, we need

$$(XI.3) \qquad \int_{\mathbb{R}} |\psi(q, t)|^2 \, dq \; < \; \infty \text{ for all } t,$$

and if we restrict attention to complex valued square-integrable functions $L^2(\mathbb{R})_C$, the initial data $\psi(q, 0)$, $-\infty < q < +\infty$, fix a unique solution of (XI.2) for all past and future times. This fixation seems to contradict the claim made in Ch. III that it is impossible to have a Galilean invariant law for a scalar quantity that allows the quantity to vary in space and that determines the future value of the quantity from its present value. The resolution of the seeming contradiction is that if the Schrödinger equation is assumed to be Galilean invariant, then ψ is not a scalar (see Sec. 2 below).

Moreover, Schrödinger time evolution preserves the L^2 norm $\| \|_2$ on states (that is, $\|\psi(t)\|_2 = \|\psi(0)\|_2$) implying stability in the past and future, in contrast to what happens in classical particle mechanics where sensitive dependence on initial conditions is often the case and where as a result precise prediction from initial data containing any error is an impossibility (see Ch. IX). And, again in contrast to classical particle mechanics, a confined quantum system (particles in a box) can never exhibit the property of mixing or any of the higher reaches of the ergodic hierarchy that in the classical domain helped to bridge micro-determinism and macro-randomness (see Ch. IX). In sum, quantum mechanics seems not only as deterministic but more deterministic, more predictable, and less stochastic than classical mechanics.

We will gradually see why these first impressions are grossly mis-leading.

2. THE QUANTUM STATE: A CLOSER LOOK

The $L^2(\mathbb{R})_{\mathbb{C}}$ functions form a concrete realization of the axioms for a Hilbert space $\mathcal{H}$, a separable linear vector space over $\mathbb{C}$ equipped with a strictly positive scalar product $(\cdot, \cdot)$.[1] It is a basic assumption of the quantum theory that a physical system S is to be described by a Hilbert space $\mathcal{H}_S$ with each observable o of S being represented by a linear self-adjoint operator $\mathbf{0}$ on $\mathcal{H}_S$. Exactly how the associations $S \mapsto \mathcal{H}_S$ and $o \mapsto \mathbf{0}$ are to be implemented is left somewhat to the creative imagination of the physicist, though some guidance is provided by concrete examples and recipes for cooking up a quantum description from a classical Hamiltonian description.

Problems arise especially for recipes for cooking up the quantum operators corresponding to products of classical observables whose operators are non-commuting. The problems are most acute for products $q^m p^n$ of classical position q and momentum p. Von Neumann (1955) proposed the following restrictions on correspondence rules:

(vN) (1) if $o \mapsto \mathbf{0}$, then $g(o) \mapsto g(\mathbf{0})$

(2) if $o \mapsto \mathbf{0}$ and $o' \mapsto \mathbf{0}'$, then
$o + o' \mapsto \mathbf{0} + \mathbf{0}'$ (regardless of whether $\mathbf{0}$ and $\mathbf{0}'$ commute).

Using the canonical commutation relations

(C) $[\mathbf{P}, \mathbf{Q}] = \mathbf{PQ} - \mathbf{QP} = -i\hbar$

for the operators $\mathbf{P}$ and $\mathbf{Q}$ corresponding respectively to p and q,[2] the (vN) rules can be shown to be inconsistent with the understanding that $\mapsto$ associates a unique $\mathbf{0}$ with each o (see Shewell (1959)).

Special cases of the von Neumann rules are essential to ordinary quantum mechanics. For instance, it is assumed that the quantum description of a simple harmonic oscillator with classical Hamiltonian $h(p, q) = p^2/2m + \frac{1}{2}aq^2$ is obtained by using $H(\mathbf{P}, \mathbf{Q}) = \mathbf{P}^2/2m + \frac{1}{2}a\mathbf{Q}^2$. Hermann Weyl's (1950) correspondence rule does give $h \mapsto H$ for this case, but it also implies that $\mathbf{0}(h^2) \neq [\mathbf{0}(h)]^2$, leading to the counterintuitive result that there should be energy dispersion even in an eigenstate of energy. A number of other correspondence rules have been proposed, e.g., Born and Jordan (1925), Dirac (1926), Rivier (1951), and Kerner and Sutcliffe (1970). Dirac's rule, like von Neumann's, is inconsistent. Among the consistent rules linearity (vN2) is generally implied while only special cases of (vN1) emerge. The observable-operator correspondence problem has been rediscovered in recent years under the label of the 'problem of hidden variables'. We will have several occasions to return to it below, but let us now turn to a closer look at the quantum state.

In what is called the 'Schrödinger picture', the instantaneous quantum state of the system S is given by a ray or one-dimensional subspace of $\mathscr{H}_S$. It is conventional to choose a normed vector ($\|\psi\| = 1$) belonging to the ray and refer to ψ as *the* state vector, any other normed ψ' belonging to the same ray being related to ψ by a constant phase factor. Once the quantum H is cooked up from the classical Hamiltonian, the change of state in the Schrödinger picture is given by

$$(XI.4) \quad i\hbar \frac{\partial \psi}{\partial t} = H\psi$$

The quantum history of the system S is a path $t \mapsto \psi(t)$ in $\mathscr{H}_S$. That agreement of histories at one point of a path, located in a Hilbert space which resides in Plato's heaven or wherever it is that Hilbert spaces reside, forces agreement on the entire path is a kind of determinism, but it is not the kind we are used to. All of the other fundamental theories of physics, be they Newtonian, special relativistic, or general relativistic, are space-time theories in that the history of a system is specified by geometric object fields on the space-time manifold,[3] and determinism for these theories means that agreement of physically possible histories on the values of the space-time quantities in one region of space-time forces agreement on the values in another region.

The $L^2(\mathbb{R})_C$ realization of $\mathscr{H}_S$ for a system S consisting of a single spinless particle seems to make contact with the space-time picture, but even in this simple case some fiddling is required. The wave function $\psi(q, t)$ is not, as might be expected, a scalar field on Newtonian space-time. The value $\psi'(q', t')$ in a new Galilean frame (q', t') is not uniquely determined by the value $\psi(q, t)$ in the old frame (q, t) and the transformation from (q, t) to (q', t'), for requiring that (XI.2) is Galilean invariant forces $\psi'(q', t')$ to depend essentially on the mass. (This fact implies Bargmann's superselection rule for mass, forbidding the superposition of states of different mass in non-relativistic quantum mechanics.[4]) Declaring m to be a scalar, ψ can be viewed as part of a composite geometric object (ψ, m).

For a composite system $S_1 + S_2$ another postulate of the quantum theory states that the appropriate Hilbert space is the tensor product $\mathscr{H}_{S_1} \otimes \mathscr{H}_{S_2}$. For two non-identical spinless particles we can realize the Hilbert space structure by $L^2(\mathbb{R} \times \mathbb{R})_{\mathbb{C}}$ functions. But when the particles interact, the composite state function $\psi_{12}(q_1, q_2, t)$ is generally not factorizable into the product $\psi_1(q_1, t) \cdot \psi_2(q_2, t)$, and so the composite wave function cannot be considered (part of) a local geometric object field on space-time.

This non-local, holistic feature of the quantum description was one of the reasons Einstein found the theory to be "fundamentally unsatisfactory." In "Quantum Mechanics and Reality" he wrote:

If one asks what, irrespective of quantum mechanics, is characteristic of the world of ideas of physics, one is first of all struck by the following: the concepts of physics relate to a real outside world, that is, ideas are established relating to things such as bodies, fields, etc., which claim a 'real existence' that is independent of the perceiving subject ... It is further characteristic of these physical objects that they are thought of as arranged in a space-time continuum. An essential aspect of this arrangement of things in physics is that they lay claim, at a certain time, to an existence independent of one another, provided these objects are 'situated in different parts of space'. Unless one makes this kind of assumption about the independence of the existence (the 'being-thus') of objects which are far apart from one another in space ... physical thinking in the familiar sense would not be possible. It is also hard to see any way of formulating and testing laws of physics unless one makes a clear distinction of this kind. (Born (1971), p. 170).

The kind of reality characterized by a space-time theory seems to be precisely what Einstein had in mind. The local geometric object fields O_i on the space-time manifold M certainly have a 'real existence' independent of perceiving subjects, frames of reference, points of view, etc. And they illustrate the 'being thus' or independence of existence of objects situated in different parts of spacetime; for if U and V are any open neighborhoods of M, the restrictions $O_i|_U$ and $O_i|_V$ of these O_i to U and V respectively are well defined, and the state on the combined region $U \cup V$ is nothing more than the 'sum' of the restricted states. But if this is the correct explication of Einstein's views, then I cannot agree with his conclusion that it is hard to see any way of formulating and testing physical laws outside of the space-time format; indeed, the laws of elementary QM and their experimental tests can be taken to constitute a counterexample. Nor is determinism necessarily undermined by the use of non-local objects, for giving the value of the two-particle wave function ψ_{12} on every pair of points on a plane of absolute simultaneity suffices to determine it for all other times. Local determinism is, however, partly undermined. Specifying the values of ψ_{12} for every pair of points either in a compact region R of a time slice t_0 or on the boundary of R for $t_0 \leqslant t \leqslant t_1$ will determine ψ_{12} for pairs of points within $R \times [t_0, t_1]$; but the 'sum' of such local determinations for a partition of t_0 need not add up to a complete determination of ψ_{12} for $t_0 < t \leqslant t_1$.

Part of Einstein's concern is addressed by relativistic quantum field theory (QFT) which describes quantum phenomena in something approaching a space-time format. The basic quantum observables are taken to be operator-valued geometric object fields on Minkowski space-time. Admittedly, some locality is lost since the quantum fields may not exist at individual space-time points except in a distributional sense and since the distributions may not restrict to a spacelike hyperplane. Nevertheless, much of the non-local holism of ordinary QM either disappears or else is rendered non-mysterious. Thus, the ψ-function is now seen to be functional of local quantum fields, and the non-local character of this functional is no more mysterious or disturbing than that of a functional of Newtonian or classical relativistic fields. Moreover, the local nature of determinism is restored, or so the field theorists hope to show. In the 'Heisenberg picture' (which is unitarily equivalent to the Schrödinger picture) the state vector is fixed and the quantum operators evolve. Thus, in this picture the closest quantum field theoretic analogue of local Laplacian determinism states that the field operators on the local sandwich S_ε of some finite thickness $\varepsilon > 0$ (see Fig. XI.1) uniquely determine the fields on the past

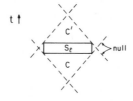

Fig. XI.1

and future caps C and C' subtended by S. This is just the 'diamond property' that local axiomatic QFT demands.[5]

Thus far I have pretended that quantum ontology is not all that different from the familiar classical world ontologies. This pretense is pierced by Born's notion that the ψ-function represents a "complete description" of reality, implying that prior to an act of observation or measurement, a quantum system may not possess sharp or determinate values for some of the observables corresponding to the quantum operators. Einstein complained to Born that, unless to the contrary, the ψ-function gives only a partial and incomplete description, quantum measurement involves ugly actions-at-a-distance. It is to the quantum measurement process that I now turn.

3. THE PROJECTION POSTULATE

Another caveat concerning Nagel's assertion that the quantum theory is deterministic derives from the Projection Postulate which appears in von Neumann's classic *Mathematical Foundations of Quantum Mechanics* and which is repeated in many standard textbooks. According to von Neumann's reformulation of the Schrödinger picture, the quantum state changes in two fundamentally different ways. When the system S is left to itself the state evolves in the smooth and deterministic manner specified by the Schrödinger equation. But when a measurement is made on S the state changes in a discontinuous and non-deterministic manner governed by the Projection Postulate. Suppose that the observable being measured has a corresponding Hilbert space operator $\mathbf{0}$ with a purely discrete spectrum. The eigenvalue equation

$$(XI.5) \quad \mathbf{0}\,\varphi_i = \lambda_i\,\varphi_i$$

determines the possible values λ_i (guaranteed to be real by the condition that $\mathbf{0}$ is self-adjoint) of the outcome of the measurement. The eigenvectors form a basis for $\mathcal{H}_S$, and if we expand the state vector at any instant in this basis

$$(XI.6) \quad \psi = \sum_i c_i \varphi_i$$

the $|c_i|^2$ give the probabilities that a measurement of $\mathbf{0}$, should it be made at that instant, would return the value λ_i. If the measurement is actually made, returning the value λ_k (say), then ψ is thrown away and replaced by φ_k. This replacement rule is unambiguous only when the spectrum of $\mathbf{0}$ is non-degenerate, with eigenvectors corresponding one-one to eigenvalues. In cases of degeneracy supplementary rules, such as that of Lüders, have been proposed; the various difficulties to which these rules are subject will not be reviewed here (see Stairs (1982) and Teller (1983)).

This replacement process is sometimes referred to as the 'collapse of the wave packet'. The name comes from picturing the probability density $|\psi(q, t)|^2$ from the $L^2(\mathbb{R})_C$ realization of Hilbert space as a wave in ordinary space. If an approximate position measurement is made, localizing the particle within some compact region R, then $\psi(q, \text{before})$ whose support may have been all of $\mathbb{R}$, collapses to a $\psi(q, \text{after})$ whose support is R (see Fig. XI.2).

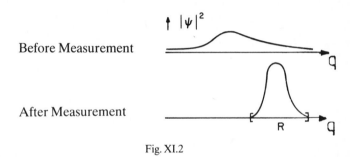

Fig. XI.2

There is nothing in the least mysterious about this collapse if the quantum probability density $|\psi(q, t)|^2$ is comparable to the ensemble density ρ of classical statistical mechanics (see Ch. IX), which also evolves in a dualistic fashion. When the system is left to itself to evolve unobserved, ρ changes in the deterministic manner given by the Liouville equation

$$(\text{XI.7}) \quad \frac{\partial \rho}{\partial t} = \{\rho, h\} \quad (\{\} = \text{Poisson bracket} \\ h = \text{classical Hamiltonian})$$

But when we make a macroscopic observation on the system the development (XI.7) is interrupted and ρ collapses to a new density function ρ' whose support is confined to the subregion of phase space compatible with the result of the observation.

For the comparison to be apt ψ must, in Einstein's terminology, represent an incomplete description of the real state of affairs. Einstein exploited examples of measurements on a composite system, first in a 1935 paper written jointly with Podolsky and Rosen (the 'EPR paradox') and later in 1948 in "Quantum Mechanics and Reality," to argue that unless the quantum description is incomplete the collapse of the wave function implies action-at-a-distance. Consider a composite system $S_1 + S_2$ consisting of two particles which are 'widely separated' in space, and let us suppose that the composite state function can be written as

$$(XI.8) \quad \psi_{12} = \sum_i c_i \varphi_{1i} \otimes \varphi_{2i}$$

where the φ_{1i} are the eigenfunctions of the operator $\mathbf{A}$ on $\mathcal{H}_{S_1}$. If we measure $\mathbf{A} \otimes \mathbf{I}$ on $S_1 + S_2$ and find (say) a_k, where $\mathbf{A}\varphi_{1i} = a_i \varphi_{1i}$, then the Projection Postulate tells us that ψ_{12} collapses to $\varphi_{1k} \otimes \varphi_{2k}$. We could have chosen instead to measure $\mathbf{B} \otimes \mathbf{I}$ where $[\mathbf{A}, \mathbf{B}] \neq 0$, and we can arrange it so that the ψ_{12} of (XI.8) can also be written as

$$(XI.9) \quad \psi_{12} = \sum_j c_j \varphi'_{1j} \otimes \varphi'_{2j}$$

where the φ'_{1j} are the eigenfunctions of $\mathbf{B}$. Finding the value b_l (say) implies a collapse to $\varphi'_{1l} \otimes \varphi'_{2l}$. The point is that whatever the measurement results a_k and b_l, the corresponding states φ_{2k} and φ'_{2l} for S_2 are never equal. But recall that S_1 and S_2 have been well separated in space so that if there is no action-at-a-distance the measurement on S_1 can affect only that part of space to which S_1 is confined and can have no direct influence in the physical reality in the remote part of space in which S_2 is confined. From what we have seen, however, the Projection Postulate implies that this is impossible if φ_{2k} and φ'_{2l} represent different "real states of affairs."

Einstein had an uncanny ability to locate weak points in a theory, and the present case is no exception. But it is not easy to say what the force of his non-locality objection is. In non-relativistic QM it is hardly disturbing to have action-at-a-distance; indeed, we have already seen that the Schrödinger equation in Newtonian space-time implies infinitely fast disturbances in the ψ-field. In relativistic QM it would be disturbing to have action-at-a-distance in the form of causal signals propagating faster than light; but an analysis of Einstein's thought experiment shows that the correlations between S_1 and S_2 cannot be used to send messages from one wing to the other. This is not the place to try to discuss further details of locality and non-locality in quantum physics.[6] But what we must do is to examine Einstein's assumption that it is an open option to grasp the second horn of his dilemma: either QM involves action-at-a-distance or else the ψ-function is an incomplete description of reality, a reality that, presumably, is recognizably similar to that of non-quantum worlds. We will now see why there must be some hesitancy in reaching for the second horn.

4. THE INCOMPLETENESS OF QUANTUM MECHANICS: JOINT PROBABILITIES

That quantum mechanics gives an "incomplete description" of reality is most obviously interpreted to mean that quantum mechanics is analogous to classical statistical mechanics where the ensemble density $\rho(p, q)$ represents our partial knowledge-partial ignorance of the exact microstate (p, q). Trying to take this analogy literally leads to a maze of problems which are worth exploring for the light and shadow they case on quantum ontology.

In the $L^2(\mathbb{R})_C$ realization of Hilbert space, the quantum state gives a probability density $|\psi(q)|^2$ for position q and, taking the Fourier transform $\phi(p)$, a probability density $|\phi(p)|^2$ for momentum p. But standard quantum mechanics gives no joint probability density for p and q corresponding to the ensemble density $\rho(p, q)$. Of course, we can trivially create the joint density

$$(XI.10) \quad \rho_{QM}(p, q) \equiv |\psi(p)|^2 \cdot |\phi(q)|^2.$$

But (XI.10), which makes p and q probabilistically independent, does not do justice to the quantum correlations coded in ψ. An initial indication of this can be gained from the fact that (XI.10) contains less information than does ψ; that is, ρ_{QM} does not always determine ψ up to a constant phase factor, as can be verified by considering cases where ψ and ψ^* are not equivalent.

To do justice to quantum correlations we would have to consider functions of p and q, but this leads us back to the still unresolved problem of the observable-operator correspondence. Nevertheless, we can obtain some results, albeit negative, without first having to solve this problem.

Let $\mu_{p,q}^{\psi}$ be a probability measure on Borel sets of $\mathbb{R} \times \mathbb{R}$. To interpret this measure as the joint quantum momentum-position measure we must certainly have:

(a)   The marginal measures $\mu_p^{\psi}(S) \equiv \mu_{p,q}^{\psi}(S \times \mathbb{R})$ and $\mu_q^{\psi}(T) \equiv \mu_{p,q}(\mathbb{R} \times T)$ agree respectively with the quantum mechanical probabilities $\text{Pr}_{\mathbf{P}}^{\psi}(S) = \int_S |\phi(p)|^2 \, dp$ and $\text{Pr}_{\mathbf{Q}}^{\psi}(T) = \int_T |\psi(q)|^2 \, dq$ for any Borel sets S and T.

Further, if f is a two-place Borel function, then we can define the probability of the phase space combination $f(p, q)$ by

$$\mu_{f(p, q)}^{\psi}(S) \equiv \mu_{p,q}^{\psi}(f^{-1}(S))$$

Thus, if $F(\mathbf{P}, \mathbf{Q})$ is the self-adjoint operator corresponding to the classical phase space observable $f(p, q)$ we should also have:

(b) $\mu_{p,q}^{\psi}(f^{-1}(S)) = \text{Pr}_{F(\mathbf{P}, \mathbf{Q})}^{\psi}(S)$ for any two-place Borel function f and any Borel set S.

Fine (1982b) shows that the combination of (a) and (b) is inconsistent with ordinary quantum mechanics in that it implies that $[\mathbf{P}, \mathbf{Q}] = 0$.

This inconsistency can be cast into what is perhaps an even more disturbing form. In the Heisenberg picture the position operators $\mathbf{Q}_t$ and $\mathbf{Q}_{t'}$, for $t \neq t'$, do not commute. Thus, under the above strictures there is no joint quantum probability for position at different times. Note also that, contrary to what is sometimes asserted, the Heisenberg uncertainty relation

(U) $\Delta p \cdot \Delta q \geq h$

where Δp and Δq are respectively the quantum standard deviations for momentum and position, does not by itself preclude the existence of a joint momentum-position distribution, as shown by the fact that (U) is entailed by the trivial joint defined in (XI.10) (see Cohen (1966a, 1966b)).

What is the origin of this impossibility result? For a particular phase function $f(p, q)$ we may well be able to choose an appropriate corresponding self-adjoint $F(\mathbf{P}, \mathbf{Q})$ such that for a classical probability measure $\mu_{p,q}^{\psi}$ we have $\mu_{p,q}^{\psi}(f^{-1}(S)) = \text{Pr}_{F(\mathbf{P}, \mathbf{Q})}^{\psi}(S)$. But we cannot do this systematically, preserving functional relations. If g is a one-place Borel function, we would expect that for the operator $G(\mathbf{P}, \mathbf{Q})$ corresponding to $g \cdot f(p, q)$ we have

$$\mu_{p,q}^{\psi}((g \cdot f)^{-1}(S)) = \Pr_{G(\mathbf{P}, \mathbf{Q})}^{\psi}(S)$$
$$= \mu_{p,q}^{\psi}(f^{-1}(g^{-1}(S)))$$
$$= \Pr_{F(\mathbf{P}, \mathbf{Q})}^{\psi}(g^{-1}(S))$$
$$= \Pr_{g(F(\mathbf{P}, \mathbf{Q}))}^{\psi}(S)$$

Since this is to hold for all quantum states, we have, if we identify operators which are probabilistically indistinguishable,

(XI.11) if $o \mapsto \mathbf{0}$, then $g(o) \mapsto g(\mathbf{0})$ for any
$o = f(p, q)$ and any one-place Borel function g,

which is the first half of von Neumann's correspondence rule. The computations of Cohen (1966a, 1966b) show how the breakdown in defining joint quantum momentum-position probabilities occurs at just this stage of trying to mirror functional relations among classical observables in the corresponding quantum operators.[7] There is an even darker side of this breakdown, as we will see in the next section.

5. THE THEOREMS OF GLEASON AND KOCHEN-SPECKER

If the correspondence $\mapsto$ between classical phase space observables and self-adjoint operators were one-one and satisfied the von Neumann rule (XI.11), then each classical micro-state (p^*, q^*) would define a real-valued valuation function $\mathrm{val}(\cdot)$ from quantum operators (in the range of $\mapsto$) such that

(V) $\mathrm{val}(g(\mathbf{0})) = g(\mathrm{val}(\mathbf{0}))$,

for we can set $\mathrm{val}(\mathbf{0}) = f(p^*, q^*)$ where $o = f(p, q)$ is the classical phase space observable corresponding to $\mathbf{0}$.

We can now ask the more general question of whether there is a valuation function, based on classical phase space or not, from some set θ of self-adjoint operators to $\mathbb{R}$ (say) satisfying (V) for all Borel functions g. There are various answers, most of them strongly negative. To see why this is so let us first note some easy but important consequences of (V).

If **A** and **B** commute, they can be written as functions of a common **C**, with the result that (V) requires the sum and product rules:

(Sum)   val(**A** + **B**) = val(**A**) + val(**B**)

for commuting **A** and **B**.

(Prod)   val(**AB**) = val(**A**) · val(**B**)

Next note how val must act on projection operators. A projection operator **E** is self-adjoint and idempotent, i.e., **EE** = **E**, with the consequence that val(**E**) must be 0 or 1. Projection operators are in one-one correspondence with the closed subspaces of $\mathscr{H}$, and for projections $\mathbf{E}_i$ onto mutually orthogonal subspaces we have from (Sum) that val($\Sigma_i \mathbf{E}_i$) = Σ_i val($\mathbf{E}_i$) for finite sums. The identity operator **I** is a projection and we assume that val(**I**) = 1, with the consequence that val($\mathbf{\Theta}$) = 0 for the null operator $\mathbf{\Theta}$.

From (V), val($\chi[\lambda](\mathbf{O})$) = $\chi[\lambda]$(val(**O**)), where $\lambda \in \mathbb{R}$ and $\chi[\lambda]$ is the characteristic function of the set $\{\lambda\}$. So for any real λ, val(**O**) = λ just in case val($\chi[\lambda](\mathbf{O})$) = 1. But $\chi[\lambda](\mathbf{O})$ is the projection operator onto the subspace spanned by the eigenvectors of **O** with eigenvalue λ. Thus we have the spectrum rule

(Spec)   val(**O**) is an eigenvalue of **O**.

Fine and Teller (1978) establish that requiring (Spec) plus (Sum) for all self-adjoint operators is equivalent to (V), as is requiring (Spec) plus (Prod).

For operators, like position and momentum, with purely continuous spectra, there are no eigenvalues so that we already have an impossibility of a valuation function satisfying (V) if θ contains such operators. So let us restrict attention to operators with purely discrete spectra. For dim($\mathscr{H}$) = 3, consider a triple α, β, γ of mutually orthogonal one-dimensional subspaces of $\mathscr{H}$ and let $\mathbf{E}_\alpha$, $\mathbf{E}_\beta$, and $\mathbf{E}_\gamma$ be the corresponding projection operators. If θ contains all such projections as α, β, γ range over all triples of orthogonal one-dimensional subspaces, then there can be no val: $\theta \to \mathbb{R}$ satisfying (V). For from the above results we know that val must assign 1 to exactly one of the projectors of each triple and 0 to the other two; but a theorem of Gleason (1957) shows that such an assignment is impossible. Kochen and Specker (1967) and Bell (1966) independently strengthened this impossibility result by showing how to find a finite subset $\theta' \subset \theta$ for which no val obeying (V) exists. For a Hilbert space of dimension greater than 3 we can, of course, prove an analogous result. The case of dimension 2 is an exception; but the previous result on the non-existence of joint probabilities for non-commuting operators continues to hold.

6. BELL'S THEOREM

A related impossibility result is due to J. S. Bell (1964). We begin with a set θ containing at least four self-adjoint operators A_i, B_j ($i, j = 1, 2$), all of which are assumed to be bivalent having eigenvalues ± 1 (say). A_1 and A_2 are non-commuting, as are B_1 and B_2, but $A_1(A_2)$ commutes with $B_1(B_2)$. In an attempt to give a classical interpretation we imagine that there is a generalized phase space Ω and that for each state $\omega \in \Omega$ there is a real valued valuation function val_ω satisfying the spectrum condition $\mathrm{val}_\omega(A_i) = \pm 1$, $\mathrm{val}_\omega(B_j) = \pm 1$. Next we assume that there is a normalized probability density ρ^ψ on Ω which returns the correct quantum probabilities for the quantum state ψ for singleton members of θ and for commuting pairs. For example, the generalized phase space probability for, say, $A_i = +1$ is

$$(\text{XI.12}) \quad \mathrm{pr}^\psi(A_i = +1) \equiv \int_\Omega \rho^\psi(\omega)\chi[A_i^+](\omega)\,d\omega$$

where $\chi[A_i^+]$ is the characteristic function of the set $\{\omega \in \Omega: \mathrm{val}_\omega(A_i) = +1\}$. And the generalized phase space probability for $A_i = +1$ and $B_j = +1$ is

$$(\text{XI.13}) \quad \mathrm{pr}^\psi(A_i = +1, B_j = +1) \equiv \int_\Omega \rho^\psi(\omega)\chi[A_i^+ \cap B_j^+](\omega)\,d\omega$$

$$= \int_\Omega \rho^\psi(\omega)\chi[A_i^+](\omega) \cdot \chi[B_j^+](\omega)\,d\omega$$

If θ is large enough we run into the Gleason—Kochen—Specker problems (Fine (1982b)). Specifically, suppose that there are $C_i \in \theta$ such that $A_i = f(C_i)$ and $B_i = g(C_i)$ for Borel functions f and g and that for any Borel set S, $\chi_S(A_i) \in \theta$ and $\chi_S(B_i) \in \theta$. Then for almost any $\omega \in \Omega$, val_ω must obey the Kochen—Specker valuation rule (V) and the product rule (Prod). But even if θ is not large enough, we are still in contradiction to the statistical predictions of quantum mechanics.

To see this, we pull out of thin air a purely number theoretic result due to Clauser and Horne (1974):

Lemma. If x, x', y, y', X, Y are real numbers such that
$$0 \leqslant x, x' \leqslant X \text{ and } 0 \leqslant y, y' \leqslant Y, \text{ then}$$
$$-XY \leqslant xy - xy' + x'y + x'y' - Yx' - Xy \leqslant 0.$$

In our application the characteristic function is bounded by 0 and $+1$ so by taking $X = Y = +1$ we have

$$(\text{XI.14}) \quad -1 \leqslant \chi[A_1^+](\omega) \cdot \chi[B_1^+](\omega) - \chi[A_1^+](\omega) \cdot \chi[B_2^+](\omega)$$
$$+ \chi[A_2^+](\omega) \cdot \chi[B_1^+](\omega) + \chi[A_2^+](\omega) \cdot \chi[B_2^+](\omega)$$
$$- \chi[A_2^+](\omega) - \chi[B_1^+](\omega) \leqslant 0$$

Multiplying by ρ^ψ and integrating over Ω we obtain

$$(\text{XI.15}) \quad -1 \leqslant \text{pr}(A_1 = +1, B_1 = +1) - \text{pr}(A_1 = +1, B_2 = +1)$$
$$+ \text{pr}(A_2 = +1, B_1 = +1) + \text{pr}(A_2 = +1, B_2 = +1)$$
$$- \text{pr}(A_2 = +1) - \text{pr}(B_1 = +1) \leqslant 0$$

The family of such relations is referred to collectively as the Bell—Clauser—Horne inequalities. There are quantum states for which the quantum probabilities are provably in violation of the BCH inequalities. Moreover, experiments confirm the quantum mechanical predictions (see Clauser and Shimony (1978) and Aspect *et al.* (1982)).

Much of the discussion of the Bell theorems has focused on the issue of locality vs. action at a distance. In the folklore the theorems are often glossed as establishing the impossibility of a local hidden variable interpretation. But, in the first instance, the issue of locality is a red herring. The version of the Bell theorem presented above is not correctly represented as establishing the conditional (locality + X) $\Rightarrow$ Bell's inequalities. The only locality used in the above derivation was semantic locality in the sense that the $\mathbf{A}_i$ and $\mathbf{B}_j$ correspond to non-relational quantities. Nor does any amount of honest Newtonian action at a distance in the form of non-contiguity or infinitely fast signals — both of which are compatible with the classical space-time structure assumed in ordinary quantum mechanics and the second of which is implied by the Schrödinger equation — block the impossibility result.

The impossibility emerges from the X — the existence of a phase space representation — whether or not Nature operates locally or at a distance. Since the phase space representation gives rise to the existence of joint probabilities for non-commuting operators, the impossibility can be seen as an expression of the quantum theory's prohibition against such joints.[8] In fact, the BCH inequalities are sufficient as well as necessary for the existence of a joint probability for the quartet A_1, A_2, B_2, B_2 returning the correct quantum probabilities for singleton operators and commuting pairs (Fine (1982a)).

The Bell theorems do have an indirect bearing on the issue of locality in the quantum domain. It is precisely because the derivation of the Bell inequalities does not require an assumption of physical locality that the Bell impossibility result is strong support for the conclusion that quantum observables do not have simultaneously sharp values. This conclusion deepens the mystery of quantum measurement. And when a sharp value emerges on one wing of an experiment as a result of a local measurement operation on a distant wing, some kind of spooky action at a distance does seem involved. The mystery of measurement will occupy us in the sections to come. It has been claimed that Bell's theorem tolls for determinism; this claim will be evaluated in Sec. 10.

7. REALISM AND THE INCOMPLETENESS OF QUANTUM MECHANICS

Recall that Einstein described his point of departure as 'realistic' rather than 'deterministic'. It is precisely realism rather than determinism which, in the first instance, is called into question. Realism (with a capital 'R') for Einstein consisted of at least three elements. (R1) In his own words, "the concepts of physics relate to a real outside world, that is, ideas are established relating to such things as bodies, fields, etc., which claim a 'real existence' that is independent of the perceiving subject ..." (Born (1971), p. 170). (R2) These 'real existents' are characterized by physical magnitudes which have simultaneously determinate values (e.g., a particle "really has a definite position and a definite momentum, even if they cannot both be ascertained by measurement in the same individual case" (Born (1971), p. 169)). Whether or not these definite or determinate values evolve deterministically is a separable issue. (R3) Statements about probabilities at a given time must be interpretable as statements about unknown but determinate values at the given time. I know of no place where Einstein explicitly states (R3), but his remarks indicate that he took it as a corollary of (R1) and (R2). In any case, it is by assuming (R3) that we can understand what he meant by the 'incompleteness' of quantum mechanics; namely, the probability statements of the theory provide partial but incomplete information about the exact but unknown values guaranteed to exist by (R1) and (R2).

The results discussed above in Sec. 5 are a direct challenge to (R2) while those reported in Secs. 4 and 6 challenge (R3) directly and, thus, (R2) indirectly. It is a pity that these results were not in hand while Einstein was alive, for his reactions to them would surely have been valuable. But rather than speculate about what his reactions might have been, I will mention some of the options that have actually been proposed.

Option 1. Maintain Einstein's Realism for the magnitudes corresponding to quantum self-adjoint operators. To maintain (R2) in the face of the Gleason—Kochen—Specker results, reject (V) as a constraint on valuation functions. We have seen that practicising quantum mechanicians do in fact reject von Neumann's condition of preservation of functional relations in moving from classical observables to quantum operators, so it should not be surprising that there are proposals to reject it going in the other direction (see Fine (1973) and Kuryshkin (1977)). We know from Sec. 5 that the price to be paid for this move is an abandonment of either the spectrum rule or else the sum and product rules. For the case of quantum operators with continuous spectra there is. obviously strong reason for the Realist to abandon (Spec). In the case of operators with discrete spectra it must be explained why a measurement yields results in conformity to (Spec) although the corresponding magnitudes (allegedly) have values other than their quantum eigenvalues. As for (Sum) and (Prod) there is disagreement about the extent to which actual experimental results confirm these rules (see Cartwright (1977), Fine (1977), Glymour (1977), and Redhead (1980)).

To maintain (R2) and (R3) in the face of Bell's theorem one could reject (Spec) or else hypothesize that the statistical predictions of quantum mechanics are not quite correct. The weight of the evidence is against the latter alternative, both in the specific experiments to test the Bell inequalities (Clauser and Shimony (1978) and Aspect *et al.* (1982)) and in the vast range of successful applications of the theory. In any case, in order to keep the discussion manageable I will continue to assume that quantum mechanics is statistically correct.

Other strategies for keeping classical Realism intact include the suggestion that we abandon classical logic in favor or some funny logic (see Putnam (1970), (1976)). The more intrepid readers are invited to explore the literature on quantum logics on their own. I would suggest that these explorers keep in mind the following questions: Using quantum logic can it be proved that the valuation functions which provably don't exist by classical logic do exist after all? Either way, what has been accomplished by switching to a funny logic?

Option 2. Maintain a modified version of (R2) by relinquishing the assumption that the valuation function must be point valued (see Fine (1971), Teller (1979)). If we allow set and interval values, we could take quantum mechanics at face value and for any state ψ define a valuation function val^ψ in the following way. If $\mathbf{O}$ has discrete spectrum, let $\mathrm{val}^\psi(\mathbf{O})$ be the set of all eigenvalues λ_i such that $\mathrm{Pr}_\mathbf{O}^\psi(\{\lambda_i\}) \neq 0$. If $\mathbf{O}$ has a continuous spectrum, discard intervals $[a, b] \subset \mathbb{R}$, $a < b$, such that $\mathrm{Pr}_\mathbf{O}^\psi([a, b]) = 0$ and take $\mathrm{val}^\psi(\mathbf{O})$ to be the closure of the remainder. Then (V) is satisfied; and *sans* Projection Postulate quantum worlds are fully deterministic, for all the magnitudes corresponding to self-adjoint operators have determinate, albeit set or interval, values and these values evolve deterministically.

One seemingly awkward consequence is that if $\mathrm{val}^\psi(\mathbf{P})$ is a compact interval, then $\mathrm{val}^\psi(\mathbf{Q})$ is the entire real line, so that if the interval value for momentum of your car is finite then the interval value for position of your car is inclusive of all of space. Part of the sting can be drawn by calculating that by the rules of quantum mechanics the probability that a hunt for your two-ton car will find it on Mars rather than in your garage is relatively small. But drawing the sting in this way brings us back to the measurement problem. When you look for your car and find it (let us suppose) safely in your garage, a change of state occurs from ψ to ψ' with the corresponding change in the value of $\mathbf{Q}$ from $\mathrm{val}^\psi(\mathbf{Q}) = \mathbb{R}$ to $\mathrm{val}^\psi(\mathbf{Q}) = $ garage. If your conscious perception plays an essential role in this process the first tenet (R1) of Realism is threatened. If, on the other hand, consciousness does not play an essential role and the measurement process is a purely physical interaction between object system and measurement apparatus, then the changes that occur during measurement should be explained by the quantum theory by plugging the appropriate interaction Hamiltonian into the Schrödinger equation. But as we will see in the next sections it is doubtful that such an explanation is forthcoming.

Option 3. Contextualist interpretations posit that it is not a quantum operator but the operator *plus* the 'context' that corresponds to an observable having a determinate value in each total state of the world. Shimony (1984) distinguishes two kinds of contextualism, environmental and algebraic. Environmental contextualism, which makes the value assignment relative to the physical environment (e.g., the experimental arrangement) adds no new possibilities since the environmental variables can be included in the total state description.[9] Algebraic contextualism, which makes the value assignment to an operator depend on other operators, arose in a self-serving manner. There is a valuation function satisfying (V) for maximal self-adjoint operators (see Gudder (1970)). Contextualism was supposed to explain why the valuation does not extend to non-maximal operators. The explanation would have some attraction if the measurement of a non-maximal operator could be achieved only via the measurement of a maximal one, but that is not the typical case in actual experiments. Contextualism would make good sense for a self-adjoint operator corresponding to a relational property, with the specification of the 'context' needed to fill in the extra argument place of the relation so as to make a genuine non-relational property. But it is implausible to view the typical operator associated with a particle as having a hidden argument place, especially if that argument place is to be filled in with another particle.

These points are illustrated by the Bell experiments, where we may take A_1 (A_2) to be the spin of particle #1 along axis $\hat{a}_1$ ($\hat{a}_2$) and B_1 (B_2) to be the spin of particle #2 along axis $\hat{b}_1$ ($\hat{b}_2$). $A_i \otimes I$ and $I \otimes B_j$ are non-maximal on $\mathscr{H}_1 \otimes \mathscr{H}_2$; but, supposing that $\mathscr{H}_1$ and $\mathscr{H}_2$ are each of dim 2, A_i and B_j are maximal on their respective spaces so that the only available 'contextual' material for either particle comes from the other particle. But we surely do not want to treat talk of the spin of a particle as elliptical for talk of a relational attribute of two or more particles.

Option 4. Maintain that there must be a level at which Einstein's Realism holds, but admit that Realism does not apply directly to all quantum magnitudes because some are non-occurrent dispositional properties and as such may not have determinate values in all total states.

If we think of observables on the model of classical phase functions $f(p, q)$ then, as already noted, each total state yields a valuation function with all observables in its domain. But this may be the wrong model for observables, classical as well as quantum. Consider the coin now in my pocket and the associated observable with values $+1$ and -1 corresponding to landing heads up or heads down on my desk. In the current total Newtonian state of the world this observable has no value, but obviously neither Realism nor determinism is thereby contradicted.

Contextualism is sometimes confused with the dispositional view of properties, but the two should be kept separate.[10] While it is implausible to take spin along a specified axis to be a relational property with a hidden argument place, it is not unreasonable to take it to be a non-relational property which is dispositional like the heads-up, heads-down property of the coin in my pocket and which may or may not have a determinate value in some total states.

Option 5. In line with Option 4 agree that quantum magnitudes do not have simultaneously determinate values, but relinquish the classical faith that there must be a base level at which Realism holds. A quantum world is then a world of irreducible propensities to display determinate values when the appropriate measurement is made. How these potentialities are actualized is a form of the problem of measurement.

8. THE PROBLEM OF MEASUREMENT

Part of the attractiveness of Einstein's Realism derives from the fact that if all three tenets (R1)—(R3) held there would be no special problem about quantum measurement; quantum measurement, like classical measurement, would be seen as the discovery of pre-existing values, and the dualistic change in the quantum state postulated by von Neumann would be the natural analogue of the dualistic change of state in classical statistical mechanics. But if we turn away from Realism, we run head-on into a problem that, I will argue, admits of no plausible solution within standard quantum mechanics.

Consider how a non-Realist would interpret the measurement of a quantum magnitude corresponding to a self-adjoint operator $\mathbf{A}$, assumed to have discrete spectrum with eigenvalues λ_i. If before measurement the state ψ of the system was not an eigenstate φ_i of $\mathbf{A}$ then for the non-Realist the system did not have a sharp value of the magnitude in question because it had a set or interval value or because it had no value at all and only a propensity to display a value upon measurement. If now the measurement is actually performed returning the value $\lambda_{@}$, the state changes from ψ to $\varphi_{@}$ in which the magnitude has a sharp value. This ontological change calls for an explanation which, presumably, should be obtained from the quantum theory itself by treating the measurement as an interaction between the object system and the measurement apparatus.

We first attempt to obtain such an explanation by imagining an ideal measurement interaction. We suppose that the measuring apparatus can be prepared in an initial state $\xi_{0,k}$ (k a degeneracy parameter) such that if the object system is in an eigenstate φ_i of $\mathbf{A}$ before the interaction is switched on, the subsequent temporal evolution of the combined object + apparatus system is

$$(XI.16) \quad \varphi_i \otimes \xi_{0,k} \xrightarrow{\Delta t} \varphi_i \otimes \xi_{i,\,l(i,\,k)}$$

where the arrow indicates Schrödinger evolution and Δt is a finite time interval. This is the first respect in which the measurement is ideal; namely, the eigenvalue λ_i in the object system is not disturbed by the measurement interaction. Further, to assure that the measuring apparatus gives unequivocal information about the object system, assume that $\xi_{i,j}$ and $\xi_{i',j'}$ are orthogonal whenever $i \neq i'$ and are macroscopically distinguishable. If you like, think of the $\xi_{i,j}$ for different values of i as pointer positions on a macroscopic dial. (For future reference, let $\mathbf{D}$ be the operator corresponding to pointer position.) (XI.16) then says that Δt seconds after the initiation of the measurement interaction the pointer position i is correlated with 100% certainty with the object state φ_i.

Now suppose that the object system initially has no sharp value of the magnitude corresponding to $\mathbf{A}$ because the initial state is a non-trivial superposition $\Sigma_i\, c_i \varphi_i$ over eigenstates of $\mathbf{A}$ with different eigenvalues. Then by the linearity of the Schrödinger equation it follows from (XI.16) that

$$(XI.17) \quad \sum_i c_i \varphi_i \otimes \xi_{o,k} \xrightarrow{\Delta t} \sum_i c_i \varphi_i \otimes \xi_{i,\,l(i,\,k)},$$

the end product being a superposition of correlated $\mathbf{A}$ eigenstates and pointer position states. If we attempt to describe this upshot consistently in the same terms used to describe the object system we are forced to say that since the final state is not an eigenstate of $\mathbf{D}$ (or more properly of $\mathbf{I} \otimes \mathbf{D}$), there is no sharp pointer position. But, of course, when we look at the dial we always see that pointer pointing in a definite direction. If that direction is n the Projection Postulate implies that the state of the composite system undergoes the non-Schrödinger transformation $\rightsquigarrow$

$$(\text{XI}.18) \sum_i c_i \varphi_i \otimes \xi_{i,\, l(i,\, k)} \rightsquigarrow \varphi_n \otimes \xi_{n,\, l(n,\, k)}.$$

Bringing in an observer *qua* physical system to 'read' the dial gives only a more elaborate version of what we already have. Using v to denote the states of the retina of the observer, with $v_{i,\, u}$ corresponding to retinal registration of the pointer in position i, equation (XI.17) is replaced by

$$(\text{XI}.19) \sum_i c_i \varphi_i \otimes \xi_{o,\, k} \otimes v_{o,\, u} \xrightarrow{\Delta t'} \sum_i c_i \varphi_i \otimes \xi_{i,\, l(i,\, k)} \otimes v_{i,\, w(i,\, u)}$$

The subsequent collapse to $\varphi_n \otimes \xi_{n,\, l(n,\, k)} \otimes v_{n,\, w(n,\, u)}$ is just as unexplained as the original one. Nor does pushing the analysis deeper into the physiology of the observer, moving from retinal to brain events, serve to do anything more than to produce a more elaborate version of the same problem.

Before going further it is well to check to make sure that the problem is not an artifact of the idealizations imposed upon the measurement process. We assumed that the measurement interaction is non-disturbing. We must be prepared to relax this assumption if there are additive conserved quantities whose operators do not commute with $\mathbf{A}$ (see Wigner (1952), Yanase (1961), and Ghirardi and Rimini (1982)). We also assumed that the interaction produced a perfect correlation between object states and pointer positions, but in practice we must be content with something less than 100% correlations. And finally we assumed that the initial state of object + apparatus was pure, corresponding to a ray of Hilbert space. But because of epistemological uncertainties the actual initial state may be a statistical mixture of pure states. With all of this in mind we can ask: When the idealizations are removed but the interaction remains a recognizably measurement interaction, is it possible that Schrödinger evolution produces a final object + apparatus state that is a mixture of pure states, in each of which the apparatus has a sharp pointer position? Since the question is vague it can never be given a definitive answer, but a series of precise negative results provides strong grounds for the conclusion that the measurement problem is not an artifact of our formulation (see Earman and Shimony (1968), Fine (1970), and Shimony (1974)).

9. THE INSOLUBILITY OF THE MEASUREMENT PROBLEM

There are almost as many reactions to the measurement problem as there are workers in the foundations of quantum mechanics. I will confine myself to a discussion of four reactions which illustrate the range of possibilites and the desperate ends to which the problem drives us.

(i) *The quantum theory is false.* If one is willing to fault the principles of the theory there are various places at which to point the finger of blame. One obvious target is the application of the superposition principle to macroscopic observables. Einstein proposed to Pauli that only a proper subclass of Hilbert space states are physically realizable; in particular, he held that a realizable ψ must give macro-objects sharply defined positions, e.g., Δq for the pointer position on the dial must be less than the dimensions of the macroscopically distinguishable divisions on the dial. This quickly leads to a super-selection rule since ψ_1 and ψ_2 may both meet the constraint while $c_1 \psi_1 + c_2 \psi_2$ ($c_1, c_2 \neq 0$) does not.

Here is Pauli's response, as reported to Born:

I believe it to be *untrue* that a 'macro-body' always has a quasi-sharply-defined position, as I cannot see any fundamental difference between micro- and macro-bodies, and as one always has to assume a portion which is indeterminate to a considerable extent whenever the *wave-aspect* of the physical object concerned manifests itself. The appearance of a definite position x_0 during the subsequent observation . . . is then regarded as being a 'creation' existing outside the laws of nature . . . The natural laws only say something about the *statistics* of these facts of observation. (Born (1971), p. 223)

I agree with the first part of Pauli's response: there are no grounds for thinking that the laws of nature respect or even recognize a sharp-micro/macro cut, and so reasons for thinking that the laws of quantum mechanics are valid for the scale of atoms are also reasons for thinking they are valid for the scale of middle sized macro-objects. But the second part of Pauli's response is a beautiful, though unintended, restatement of what is so disturbing about quantum measurement: the appearance of the definite value is a "'creation' existing outside the laws of nature" not because these laws are statistical but because the 'creation' apparently violates the laws.

Another place at which to direct the blame is at the dynamics of the theory; more specifically, it is the linearity of the Schrödinger equation that leads to (XI.16) and (XI.19). There are proposals for a non-linear replacement for the Schrödinger equation designed so that a super-position is ground into an appropriate eigenstate (see Pearle (1976)). While this may prove to be the correct response I will not pursue it here. For although the discussion of these matters is necessarily speculative, it seems wisest to tie the speculations as closely as possible to the orthodoxies of the theory; otherwise we open a Pandora's Box of uncontrolled possibilites.

(ii) *Mentalism.* The second category of reaction could be classified as a subcategory of the first, but because of its startling consequences it deserves a separate treatment. We saw that tracing the process of observation ever deeper into the physiology of the observer serves only to compound the mystery of measurement. What we did not reckon with is the possibility that it is the act of registration on the conscious-ness of the observer that is responsible for the reduction of the superposition and the creation of the sharp value. This possibility, first enunciated by von Neumann, was explored by London and Bauer (1939) and more recently by Wigner (1961). I am not enough of a materialist to be convinced that a Cartesian mental-physical dualism is beyond the pale, but like Einstein I balk at the notion of inventing a causal mechanism according to which the conscious looking at the dial creates a definite pointer position where none existed before (see Born (1971), p. 222). However, I want to register my admiration for Wigner's courageous statement of the fact that if (i) is rejected then (ii) must be taken seriously. There are other proposals for trying to wiggle out of the problem, but as we will see, they come to no more than sleight of hand.

(iii) *Wash out.* There are attempts to show that as a result of interaction with a macroscopic apparatus the phase relations in the pure-state superposition in the micro-system "wash out" (see Daneri, Loinger, and Prosperi (1962)). What is actually shown is that the final object + apparatus state approximates in some sense the sought-after mixture. But here a miss by an inch is as good as a miss by a mile. In theory, the phase relations of the superposition persist until the ultimate, and still unexplained, reduction; and it is only by taking the theory seriously in the first place that we get a problem of measurement. We can modify the theory by stipulating that various self-adjoint operators do not correspond to magnitudes that are observable even in principle and, consequently, that the theoretical difference between the sought after mixture and the state entailed by Schrödinger evolution disappears; but such a move represents a retreat to category (i).

(iv) *Deoccamization.* Everett (1957) and Wheeler (1957) proposed an interpretation of quantum mechanics which promised to provide a purely quantum theoretical explanation of measurement because the Projection Postulate is not invoked and the Schrödinger equation is never violated. A number of variations of the proposal have appeared under the label of the 'many world interpretation' (see DeWitt and Graham (1973)). What is common to all of these proposals is the idea that a measurement culminates not in a collapse in which all of the elements of the superposition save one are destroyed; rather the universe 'branches' into separate and non-interacting parts, each of which corresponds to a term in the superposition. Thus, in the situation described conventionally by (XI.19), branch #1 will correspond to the observer seeing the pointer in position 1, indicating that **A** has a value λ_1; branch #2 will consist of the observer seeing the pointer in position #2, indicating that **A** has a value λ_2; etc.

It is important to realize that this talk of 'branching' must be taken literally; the different branches must represent simultaneously real situations and not merely unactualized possibilities, else the talk of 'many worlds' is just a metaphorical way of redescribing the original problem. What is rarely explored is the implication for space-time structure of taking this deoccamization seriously. To make sure that the different branches cannot interact even in principle they must be made to lie on sheets of space-time that are topologically disconnected after measurement, implying a splitting of space-time something like that illustrated in Fig. XI.3. I do not balk at giving up the notion, held sacred until now, that space-time is a Haudorff manifold. But I do balk at trying to invent a causal mechanism by which a measurement of the spin of an electron causes a global bifurcation of space-time.

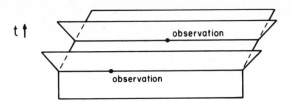

Fig. XI.3

Further, there is no principled answer to the question of when the splitting occurs. It cannot be taken to occur at the moment at which the correlation is established between object and apparatus, for there are experimentally verifiable cases where correlations are established between two systems and yet the interference terms between the different 'branches' of the superposition for the composite system remain. Nor is it fair to say that the branching takes place when the correlation is irreversibly recorded by a macroscopic memory device; for we have rejected the notion of a sharp micro/macro cut, and in most cases of quantum measurement the relevant micro-laws are time reversal invariant. This leads to the related criticism of the many worlds approach for a lack of a principled explanation of why the time reverse of Fig. XI.3 never occurs to yield an anti-measurement process where the different branches coalesce to form a superposition.

Finally, we should remind ourselves that a vector is a vector is a vector. Any vector can be represented in many different ways as a linear sum of basis vectors; in particular, there may be many alternative choices of bases for the Hilbert space of the composite object + apparatus system relative to which the final state vector can be written as linear sum of correlated object-apparatus states. The objection is not that we have a further multiplication of worlds for each choice of basis — after all, once we have started to play the multiplication game we may as well play it consistently — but that in some expansions of the state vector the sum is over terms $\varphi_i' \otimes \xi_i'$ where the ξ_i' are superpositions of pointer positions. What is needed is an explanation of why nature apparently obeys a selection principle that forbids the realization of such branches, but the many worlds interpretation is no better at providing this explanation than the more orthodox approaches are at providing an explanation of the collapse of the superposition.[11]

While I have not done justice to the nuances and subtleties of various treatments of the measurement problem, I hope to have said enough to indicate that no matter how we twist and turn, choosing one 'solution' over another is just a matter of trading one form of *deus ex machina* for another.

10. DETERMINISM AND QUANTUM MECHANICS

What is the bearing of the various no hidden variable theorems on the issue of determinism in the quantum domain? The answer, I believe, is far from evident. In what follows I will review some of the attempts to make the connection.

(i) *Determinism and determinateness.* We could argue that determinism fails in quantum worlds because: the no hidden variable theorems show that quantum magnitudes do not have simultaneously determinate and sharp values, but determinism presupposes that they do (see Glymour (1971)). Accepting, *arguendo*, the first premise, I wish to fault the second. Determinism does not presuppose sharpness of values, for we can understand determinism as a doctrine about the evolution of set or interval valued magnitudes as well as about point valued magnitudes. Determinism does seem to presuppose some minimal amount of determinateness; if the world were entirely a froth of potentialities with no magnitudes having determinate values, point or interval, one would be at a loss to say whether determinism held or failed. But quantum determinism surely does not require that all quantum magnitudes always have determinate values, for a similar requirement would falsify classical determinism.

(ii) *Tolling Bell's theorem.* To toll Bell's theorem for determinism we can argue as follows. If determinism reigns, then based on the total state ω there must be a unique outcome for the measurement of any observable. So in state ω assign as the value of $\mathbf{O}$ the uniquely determined outcome that would emerge if $\mathbf{O}$ were measured. Then assuming that the probabilities for outcomes replicate the quantum probabilities leads to a contradiction, for we can repeat the derivation from Sec. 6 above of the BCH inequalities. Hence, determinism is defeated by Bell's theorem and the confirmation of the violation of the Bell inequalities in correlation experiments.

The argument is fallacious. It is true that in any instance where $\mathbf{O}$ is measured, determinism must supply a unique outcome. It does not follow that determinism always supplies an answer to what value of **O** would emerge if a measurement of **O** were made; that is, if the actual state ω of the world is incompatible with an **O** measurement, the subjunctive conditional "If **O** were measured, the obtained value would be λ" may have no truth value for any λ. What I am proposing is not a peculiarity of quantum mechanics. In any situation in which the coin now in my pocket is flipped so that it lands on my desk, the total Newtonian state of the world determines whether it lands head up or heads down on my desk. But relative to the current actual state, which is physically incompatible with the coin's now being flipped, there may be no determinate truth value to "If the coin were flipped, it would land heads up." In terms of a phase space representation the point is that the valuation function for **O** will be a partial function. In computing the phase space probability that **O** takes a given value we need to normalize by dividing by the measure of the set of phase points where the valuation function for **O** is defined, and since the normalization factors will generally be different for different observables, the BCH inequalities need not emerge. Fine (1982c) constructs models of this type for the Bell correlation experiments.

(iii) *Dynamics*. Determinism is a doctrine about the temporal evolution of the world. It is therefore peculiar (and suspicious) that most of the putative refutations of determinism in quantum worlds ignore dynamics. An argument due to von Neumann is an exception. According to Wigner (1970), von Neumann's real but unpublished objection to hidden variable interpretations of quantum mechanics derived from a consideration of what happens in a sequence of measurements of, say, spin in various directions for a spin $\frac{1}{2}$ particle. Von Neumann felt that the successive measurement of different spin components, corresponding to non-commuting operators, would progressively restrict the compatible range of the hidden parameters until eventually there would be a high probability that the spin components would have a definite sign in all directions, contradicting the quantum statistical predictions.

This argument is especially curious in view of von Neumann's seminal contributions to modern ergodic theory. From Ch. IX we know that it is possible in principle to reach into the ergodic hierarchy to design a measure preserving flow ϕ_t on the hidden variable space so that ϕ_t loses information rapidly enough on a coarse grained level of description that successive measurement results do not have the restricting effect envisioned in von Neumann's argument.

Nevertheless, I believe that von Neumann is correct in that the fate of determinism in the quantum domain will be settled by reference to dynamics, a matter that is virtually untouched by past and present work on hidden variables. The defender of determinism has a particularly difficult task in this regard, especially if von Neumann's projection postulate is to be accommodated. To preserve the valid features of quantum dynamics, the deterministic flow ϕ_t must mirror Schrödinger evolution; but to accommodate the projection postulate, ϕ_t must depart from Schrödinger evolution when a measurement is made. From the preceding section we know that the quantum theory itself offers no consistent guidelines on how to characterize the latter set of cases.

While we await general results, positive and negative, on the compatibility of a deterministic dynamics with quantum mechanics, we can try to settle special cases. The next argument attempts to rule out a form of local determinism by tolling Bell's theorem.

(iv) *More Bell.* Although it is false that determinism *per se* implies the Bell inequalities, it is open that the implication holds for special forms of determinism and hence that some forms of determinism are refuted by the experimental confirmation of the violation of the inequalities by quantum statistics. An argument to this effect has been provided by Hellman (1982).

The first step in the argument is to repeat the derivation of the Clauser, Horne, Shimony, Holt (1969) version of the inequalities from the demand of 'counterfactual definiteness'. To explain this demand we suppose that on each wing of the Bell correlation experiment there is a measuring device that admits of two knob settings. The settings correspond to internal states of the measuring devices appropriate to measuring A_1 or A_2 on the left and B_1 or B_2 on the right. Now suppose that in fact A_2, B_1 were measured, returning the results +1, −1. Counterfactual definiteness demands, for instance, that had the knob setting on the left been such that A_1 instead of A_2 was measured, the outcome of measuring B_1 on the right would still have been −1. A construction due to Stapp (1971) and Eberhard (1977) then shows that the CHSH inequalities follow on the assumption that the frequency counts in a large number of measurements converge to their true values.

Letting α_j^k and β_j^k stand for the response of the A and B-measurements on the jth trial when the knob setting is in position k, the observed correlation coefficients in N trials will be

(XI.20) $\quad C^{(1, 1)} \equiv \dfrac{1}{N} \sum_j \alpha_j^1 \beta_j^1$

$$C^{(1, 2)} \equiv \dfrac{1}{N} \sum_j \alpha_j^1 \beta_j^2$$

$$C^{(2, 1)} \equiv \dfrac{1}{N} \sum_j \alpha_j^2 \beta_j^1$$

$$C^{(2, 2)} \equiv \dfrac{1}{N} \sum_j \alpha_j^2 \beta_j^2$$

It is the assumption of counterfactual definiteness that allows, for example, the outcome on the right to be denoted by β_j^k rather than by $\beta_j^{k,\,l}$ where the extra parameter l refers to the knob setting on the left. It is easy to see that

(XI.21) $\quad \alpha_j^1 \beta_j^1 + \alpha_j^1 \beta_j^2 + \alpha_j^2 \beta_j^1 - \alpha_j^2 \beta_j^2 \equiv \pm 2$

with the result that

(XI.22) $\quad \left| C^{(1, 1)} + C^{(1, 2)} + C^{(2, 1)} - C^{(2, 2)} \right| \leqslant 2.$

If for large N the observed correlation coefficients are equal to their theoretical values, then quantum mechanics is provably in violation of (XI.22).

The second step in the argument is an attempt to show that local relativistic determinism entails counterfactual definiteness. Recall from Ch. IV that in the context of a relativistic space-time setting, local Laplacian determinism means that the state in a region R is determined by the state on any spacelike S such that $S \subset C^-(R)$ and $R \subset D^+(S)$. Such a determinism is presumably inconsistent with direct action at a distance; certainly, as Hellman notes, local determinism implies that the outcome on one wing cannot be changed *merely* by changing the knob setting on the other wing. But more is needed to secure counterfactual definiteness. If complete determinism reigns and the settings on the measuring instruments are determined along with everything else, then in the counterfactual scenario where the setting on the A-measuring device is different from the actual setting, the state on any slice (e.g., S_1 in Fig. XI.4) through the past cone of the A-measuring event must be different from the actual state, and this latter difference can in principle eventuate in a different outcome for the B-measurement.

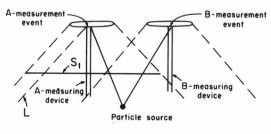

Fig. XI.4

Such an eventuality can perhaps be made implausible by specializing the hardware of the experiment. Thus, suppose that the A-measuring device is rigged so that the setting at the measurement event in question turns on causal chains that never enter the past cone of the B-measurement event; for example, the setting could be made to vary with the frequency of a source-free light ray L (see Fig. XI.4) from a direction in space chosen so that the ray is disjoint from the past cone of the B-measuring event. Then, plausibly, even if the setting on the A-measuring device had been different (because the frequency of L was different) the state on any slice in the past cone of the B-measurement would still have been the same and hence by local determinism the outcome of the B-measurement would still have been -1.

It should be emphasized that this argument does not tell against determinism *per se* but only against local determinism; for as we saw in Ch. IV, relativity theory does not require that determinism be of the local variety — at least, there are non-trivial Lorentz invariant theories which are Laplacian deterministic but not locally so. To constitute a general argument against determinism, additional evidence would have to be marshalled to show that Nature operates by contiguous action.

My main qualm about the entire line of argumentation is that it is more than slightly obscene to make ultimate judgments about the truth of such a noble doctrine as determinism turn on something as slippery as counterfactuals. If in the end the debate about whether or not the world is deterministic comes down to trading conflicting intuitions about nearness of possible worlds, then leave me out of it.

11. INDETERMINISM, RANDOMNESS, AND STOCHASTICITY

We could, if we so desired, define 'determinism' and 'chance' so that the negation of the first is equivalent to the second. But such a definition should not be allowed to obscure the fact that a failure of determinism need not be due to an irreducible stochastic element in nature. Indeed, not one of the threats to Laplacian determinism in Newtonian, special relativistic or general relativistic physics studied respectively in Chs. III, IV, and X involved probability considerations. This may, perhaps, lead some to see those threats as less interesting than the ones posed by quantum physics. But if so, it makes the former threats less easy to blunt; for it seems very unlikely that the introduction of 'hidden variables' in Newtonian gravitational theory or the general relativistic theory of gravitational collapse would serve to restore determinism without also radically altering the predictions of these theories. But as we have seen in this chapter the goal of establishing the parallel conclusion for QM has proven to be elusive.

The implication in the other direction is more appealing, for, presumably, an irreducible randomness or stochasticity would entail indeterminism. The reason for the weasel word is that there are two ways to read 'irreducible', one of which is not wanted, the other of which is difficult to understand. Empiricists cannot tolerate irreducible probabilities construed as non-Humean powers that float free of occurrent actualities. In Ch. IX we saw how the probabilities of classical statistical mechanics receive an Empiricist grounding; but that grounding, which relies on deterministic laws, undercuts the possibility we are now seeking to understand. The quantum theory holds a glimmer of a promise of a grounding for probabilities which satisfies Empiricist strictures but which does not rely on determinism. The axioms of QM combine probabilistic and non-probabilistic assumptions in an insepa-rable cluster, making it possible to have our cake and eat it too: the axioms of QM may count as the best overall deductive system for the non-probabilistic, occurrent actualities of the quantum world (e.g., the energy levels of the hydrogen atom) and thus qualify as Empiricist laws in accord with the account of Ch. V, and at the same time the axioms can have probabilistic consequences.

Even if this is correct, it still remains to ask whether quantum probabilities are 'irreducible' on a second reading of that term. In trying to understand this second reading it is useful to use classical statistical mechanics as a foil. The probabilities in classical statistical mechanics are not irreducible in that they have an *ignorance interpretation* in both an atemporal and a temporal sense. First, for any property that can be represented as a function $f(p, q)$ of phase (p, q), any probability assertion of the form 'The probability that at t the system has the property P is r $(r \neq 0, 1)$' is, in part, an expression of ignorance of the micro-state $s(t) = (p(t), q(t))$ at t; for conditionalizing on the information $s(t)$ reduces r to 0 or 1. Second, any conditional probability assertion of the form 'The probability that the system will have property Q at t' given that it has property P at $t < t'$ is r $(r \neq 0, 1)$' is also, in part, an expression of ignorance; for since the evolution of the micro-state is deterministic, the conditional probability of $s(t')$ given $s(t)$ is either 0 or 1, and combining this with the first point, the probability of Q at t' given $s(t)$ is either 0 or 1.

Irreducible stochasticity would, presumably, involve a failure of one or both of these features. On Born's interpretation of QM, quantum probabilities do not admit an ignorance interpretation of the first kind; if the quantum state at time t is such that $Pr_A^\psi(\{\lambda\}) \neq 1$ for any λ, then according to Born the system does not have at t a definite value of **A** but only a collection of propensities of strengths $Pr_A^\psi(\{\lambda\})$ to display values λ upon measurement of **A**. On one hand, various "no hidden variable" results tend to support Born's interpretation; but on the other hand, we found no coherent account of how these potentialities are actualized in the measurement process. Nor does QM provide unproblematic examples of stochasticity of the second kind. Taking QM at face value we could assume that $\psi(t)$ provides the most complete description possible of the state at t. But in general the quantum transition probabilities from the state $\psi(t)$ to later alternative states are not representable as conditional probabilities. Rather these probabilities have to be seen as propensities for the system to undergo a transition from potentialities to actualities, and again we have no coherent account of this transition. In sum, while irreducible stochasticity may be an idea whose time may come, it is far from clear that QM marks its debut.

12. CONCLUSION

Previous chapters have detailed a number of ways in which Laplacian determinism and its close relatives can fail, but not one of those cases conformed to the standard philosophical picture of indeterminism rooted in an irreducible chance or stochastic element. Quantum physics promised to bring that picture into focus, but today, over three quarters of a century since the advent of the quantum theory, we have only a blurred image. An astounding — and frustrating — feature of the theory lies in the contrast between the exquisite accuracy of its empirical predictions on one hand and the zaniness of its metaphysical 'consequences' on the other. The theory has been used to 'prove' not only that determinism is false but that realism fails, that logic is non-classical, that there is a Cartesian mental-physical dualism, that the world has the structure of Borges' garden of forking paths, etc. One is tempted to say that any theory which proves all of this proves nothing. But the temptation must be resisted. Although it is not clear what the quantum theory implies about determinism, it is clear that the implications are potentially profound. Bringing the implications into sharper focus requires a simultaneous focusing of a host of other foundations issues, most especially concerning the nature of quantum magnitudes and the nature of the quantum measurement process. By now it is no surprise that pressing the question of determinism has helped to unearth the deepest and most difficult problems that challenge our understanding of the theory.

NOTES

[1] $(\cdot, \cdot)$ is a mapping from $\mathscr{H} \times \mathscr{H}$ to $\mathbb{C}$. That it is strictly positive means that for any non-null $\psi \in \mathscr{H}$, $\| \psi \| = [(\psi, \psi)]^{\frac{1}{2}} > 0$. For the $L^2(\mathbb{R})_\mathbb{C}$ realization,

$$(\psi, \varphi) = \int_{\mathbb{R}} \psi^*(q) \varphi(q) \, \mathrm{d}q$$

The separability of $\mathscr{H}$ means that it has a countable basis. It is also assumed that $\mathscr{H}$ is complete with respect to $\| \, \|$.

[2] Unbounded operators, like $\mathbf{P}$ and $\mathbf{Q}$, are not defined on all of $\mathscr{H}$; thus, equations like (C) require a specification of a domain of definition. For the technical details, see Jauch (1968).

[3] In coordinate language, an *object field* F on a manifold M is a rule which assigns to each point $p \in M$ and each local coordinate chart x^i about p an N-tuple of numbers

$(F_1, F_2, \ldots, F_N)$ called the components of F at p with respect to the coordinate system x^i. The class of *invariant* or *geometric objects* is singled out by the requirement that if $x'^j = f_j(x^i)$ is another coordinate chart about p, then the new components F'_n of F at p with respect to the new coordinates are a well defined function of the old components F_m, the old coordinates, and the functions f_j relating the new and old charts. See Schouten (1954) for details. Familiar examples of linear geometric objects, such as vectors and tensors, can be given intrinsic, coordinate-free characterizations in terms of multilinear mappings. For one approach to a coordinate-free characterization of geometric objects in general, see Salvioli (1972).

[4] For a derivation, see Kaempffer (1965). The Φ field of the Klein-Gordon equation in relativistic wave mechanics is a genuine scalar field (its transformation from one Lorentz chart to another does not involve the mass of the particle). The Dirac spinor function *plus* a 'spin frame' together constitute a geometric object.

[5] For an overview of these matters and references to the literature, see Streater (1975).

[6] See my (1985) for a discussion of forms of locality and non-locality in classical and quantum theories. On the view Einstein is criticizing, quantum measurement involves a literal miracle — a contravention of laws (see Secs. 8 and 9 below). This is criticism enough, and it tends to moot questions of non-locality in quantum measurement. If we allow a local miracle in an otherwise local non-quantum theory, e.g., local creation of charge in Maxwell's theory of electromagnetism, dramatic non-local effects can result, e.g., the electric and magnetic fields must instantaneously adjust over all space.

[7] This reverses the actual history since it was partly by reflecting on Cohen's results that Fine was led to impose condition (b) (private communication from A. Fine). Also, the point made here does not succeed in locating the entire source of the impossibility; for the impossibility holds for Hilbert spaces of dim 2 where the functional relations can be preserved (see Sec. 5).

[8] Fine (1982b) takes the non-existence of joint probabilities for non-commuting observables to be the core of most of the no hidden variable results; but see Shimony (1984).

[9] See Shimony (1984) for a discussion of the bearing of the Bell theorems on contextualist hidden variable theories.

[10] The dispositional view may be appropriate even for the case of dim 2 where algebraic contextualist theories collapse since then all operators are maximal.

[11] For recent reappraisals of the Everett-Wheeler view see Healey (1984), Geroch (1984) and Stein (1984).

SUGGESTED READING FOR CHAPTER XI

Two good surveys of foundations problems in quantum mechanics are Jammer (1974) *The Philosophy of Quantum Mechanics* and D'Espagnat (1976) *Conceptual Foundations of Quantum Mechanics*. Belinfante's (1973) *A Survey of Hidden Variable Theories* is an excellent introduction to that topic. Reprints of many of the classic papers on the problem of quantum measurement are collected by Wheeler and Zurek (1983) *Quantum Theory and Measurement*. Elaborations of the Everett-Wheeler view are to be found in DeWitt and Graham (1973) *The Many-Worlds Interpretation of Quantum Mechanics*.

CHAPTER XII

DETERMINISM AND FREE WILL

> Over the years I have spent more time thinking about
> the problem of free will — it felt like banging my head
> against a wall — than any other philosophical topic
> except perhaps the foundations of ethics. Fresh ideas
> would come frequently, soon afterwards to curdle.
>
> (Robert Nozick, *Philosophical Explanations*)

> . . . it is really one of the greatest scandals of
> philosophy . . .
>
> (Moritz Schlick, "When Is a Man Responsible?")

The determinism-free will controversy has all of the earmarks of a dead
problem. The positions are well staked out and the opponents manning
them stare at one another in mutual incomprehension. No advances
in philosophy of science or cognitive psychology seem to move the
problem forward. (When I began writing this book I entertained the
hope that getting a more precise fix on determinism would help. I now
see how naive and vain my hope was.) Genuinely new ideas are scarce.
Some of the most intriguing ones to emerge in recent years are to found
in Robert Nozick's account of "contra-causal [read: contra-deterministic]
freedom"; but, I am sorry to report, the account is ultimately inscrut-
able. Finally, and, worst of all, for those of us who are not attached to
one of the standard positions, it has become hard to sort new proposals
into the serious and the nutty.

If the problem keeps leading us up blind alleys, why can't we walk
away from it? Because the issues it joins are essential to an understanding of human action and of man's place in nature. Here we have a
major clue as to why the problem has proved so divisive and why it
resists any neat resolution. And we also have an early indication that
'the determinism-free will problem' is a misnomer: though the apparent
conflict between determinism and free will helps to focus our attention
on puzzles about human action, the puzzles go for beyond anything to
do with determinism and free will *per se*.

I have no new solutions to offer. Nor do I have a magic Ariadne
thread that will guide us painlessly through the maze of claims and

counterclaims. But I do hope to be able to illuminate the role determinism plays in generating the controversies, and secondarily I hope to
show why thinking about these controversies is like banging your head
against the wall.

1. MORAL AND LEGAL RESPONSIBILITY AND THE COMPATIBILIST POSITION

The philosophical discussion of the determinism-free will problem has
been badly skewed by the tendency to locate it within the context of
questions about moral and legal responsibility. This context produces
an almost irresistible pressure for a compatibilist solution, a pressure
exploited by Moritz Schlick and his followers. My complaint is not that
Schlick's solution is incorrect; indeed, I think that Schlick is roughly
right about the sense of free will relevant to ordinary ascriptions of
responsibility. Rather, my objection is that Schlick's solution is shallow
and comes to grips with only a small piece of the larger problem.

Let us begin by trying to approach questions of responsibility, praise,
and blame with a vision unclouded by philosophical argumentation. If
we can locate informants innocent of philosophy, we will have little
trouble in securing commonsensical agreement on two points:

(C1)   People should be held responsible for their action unless
        there are exculpating circumstances.

(C2)   Circumstances which make it impossible for a person to act
        freely are exculpating.

If we succeed in introducing the subject of determinism in a non-
corrupting manner (one may have doubts after the preceding eleven
chapters!), we will probably also elicit:

(C3)   Discoveries in physics and the life sciences indicating that
        determinism applies to people as well as inanimate particles
        would not undermine ascriptions of responsibility.

Finding innocents ready to volunteer (C3) may not be easy, as wit-
nessed by the recent *New Yorker* cartoon (see facing page). But the fact
that we are supposed to see the humor is evidence that (C3) is the
correct commonsensical response. (Of course, there are philosophers
for whom this is no laughing matter, but remember we are trying to

proceed pre-analytically.) Now that we have elicited the responses we want, we can whip out our philosophical logic and derive the conclusion that acting freely, in so far as it is a necessary condition for responsibility, is compatible with determinism. What then is this sense of freedom? A little further prompting will produce:

(C4) Freedom is the absence of compulsion, coercion, constraint.

For it is just these factors — compulsion, coercion, and constraint — which would be regarded by the real-life counterpart of the *New Yorker* judge as being exculpating.

The title of the chapter in which Schlick discusses the free will problem — "When Is a Man Responsible?" — is a giveaway as to his

"Not guilty by reason of genetic determinism, Your Honor."

Drawing by Mankoff; © 1982.
The New Yorker Magazine, Inc.

analysis of freedom. A man is free for Schlick just in case he does not
act under compulsion, coercion, or constraint. These factors apply just
when a man is prevented from realizing his natural desires. Freedom as
the absence of these factors is quite compatible with determinism since
natural laws do not entail them — laws of nature do not *pre*scribe what
happens but only *de*scribe it. Punishment is justified as an "educative
measure," serving to "prevent the wrongdoer from repeating the act
(reformation) and in part to prevent others from committing a similar
act (intimidation)." This fits nicely with the analysis of freedom since it
is just in those cases where freedom is lacking because of compulsion,
coercion, or constraint that punishment is ineffective either as a re-
formative or intimidative measure.

Schlick took all of this to be so obvious that he felt that he had to
apologize for writing the chapter:

> With hesitation and reluctance I prepare to add this chapter . . . it is really one of the
> greatest scandals of philosophy that again and again so much paper and printer's ink is
> devoted to this matter, to say nothing of the expenditure of thought, which could have
> been applied to more important problems . . . (1966, p. 54)

But despite Schlick's attempts at intimidation there are concerns that
even our philosophical innocents will want to raise with not much more
prompting than was used to get (C1)—(C4). For example, there are
many different kinds of cases where the laws of nature and the relevant
physical circumstances combine to make it physically impossible for an
agent to perform an action A. Schlick directs our attention to cases
where the agent has formed the intention or desire to do A but
something, so to speak, comes between the agent's natural desire and its
realization. But consider cases where it is physically impossible for the
agent to, say, raise his right arm not because his arm is strapped down
or because the tug of gravity is too strong but because the laws and the
antecedent circumstances make it physically impossible for him to
desire, intend, or will to raise his arm — his arm can go up in a
convulsive twitch or jerk, but *he* cannot raise his arm. I take it that what
Schlick would say about this case would depend on the further details.
If, for instance, what has made it impossible for the agent to desire or
will to raise his arm was torture, brainwashing, or the injection of mind-
altering chemicals, then by a natural extension of Schlick's doctrine we
have a case of coercion or compulsion and, hence a case where the
agent is not free. If, on the other hand, what induces the impossibility is

no such warping of the agents 'natural' desires but the unfolding of his 'normal' genetic heritage, then Schlick would presumably say that the agent is free to raise his arm. The odd ring to this result is in no way dispelled by repeating the incantation that natural laws do not compel, they describe not prescribe. Natural laws do not compel or prescribe in the cases where it is physically impossible for the agent to raise his arm because of the strength of the gravitational field or the strength of the straps that bind his wrists, but the impossibility is no less for that. Libertarians are incensed, and rightly so, at Schlick's accusation that their bellyache is to be diagnosed as the result of swallowing the mistaken equation 'determinism by natural laws is compulsion'; rather, their bellyache is that if determinism, whatever its ontological strength, negates freedom and undermines responsibility in one type of case, then it seems to do so in others as well. That the man-in-the-street initially agrees to sort cases along Schlick's lines when it comes to answering "When is a man responsible?" hardly settles the matter. My experience is that when the man-in-the-street comes to believe that determinism applies not just to eye color and general personality traits (as the phrase 'genetic determinism' suggests) but also to the most intimate details of our outer actions and inner mental lives, then the laughter at the *New Yorker* cartoon turns nervous. And the nervousness, once induced, affects much more than questions about guilt and punishment. What is threatened is nothing less than the basis of the moral perspective and our sense of worth as human beings.

2. TROPISMS

Let us attempt to understand more fully from the Libertarian's point of view the threat that determinism poses. The Compatibilist should be interested in this goal, if only to better refute the Libertarian. To state the issues in as neutral a way as possible, let us drop for the moment questions about responsibility, guilt, and punishment and begin instead by asking questions like: How are the actions of man different from those of a sunflower as it turns to face the sun? If determinism is true, aren't all of our actions merely complicated cases of tropisms, forced motions produced by circumstances beyond our control?

The Libertarian may pose such questions in a rhetorical mode, but I propose to take them literally. The immediate and obvious difficulty then is that human actions are mediated by mental states — beliefs, desires, etc., — and mental acts — willings, choosings, etc. — so that any serious answers to the question must come to grips with the mind-body problem, a problem second only to the determinism-free will problem in its divisiveness and intractability. I claim, however, that it is possible to appreciate the church of determinism without first having to solve the mind-body problem.

We have two cases to consider: either the mental is parasitic on the physical or not. By parasitism I mean:

(P)   For any physically possible worlds W_1 and W_2, if W_1 and W_2 agree on all physical attributes then they agree on all mental attributes as well.

Parasitism *might* be explained by trying to identify mental events with physical events, but whether such an identity relation is implied by parasitism and, if so, what form the identity takes (e.g., type-type or token-token) need not be settled here. Likewise, the failure of parasitism *might* be explained by Cartesian dualism, but again the exact form of the explanation is irrelevant for our purposes. Now suppose that futuristic physical determinism holds and that if we go back far enough in time (say, to $t = -10^{10}$ years) we reach a state of the world which can be characterized in purely physicalistic terms (say, positions and velocities of particles). If (P) fails then the physical state at $t = -10^{10}$ years may not uniquely determine your current beliefs, desires, and willings. But from the assumption of futuristic physical determinism it follows that in so far as your inner mental life is autonomous it is inefficacious in producing physical actions. Imagine, if you will, that you are agonizing over the decision of whether or not to pull the trigger of a S & W .357 magnum you have trained on a mass murderer. You know that if you don't pull the trigger he is sure to escape to kill again and again; but at the same time you are revolted at the thought of taking a human life and more than a little perturbed at the thought of splattering gore over your favorite Shirvan prayer rug. Some or all of these internal debates may be underdetermined by the physical state at $t = -10^{10}$ years, but whatever differences are allowed make for no difference in the upshot since the state at $t = -10^{10}$ already determines that you do not pull the trigger (you wimp!). Alternatively, if (P) holds then not only the upshot but all of your internal debate as well is uniquely determined by the state at $t = -10^{10}$ years. Your inner deliberations combine with your outer actions to form one elaborate tropism.

Various forms of this tracing-back-the-state construction are found over and over again in Libertarian tracts. The commonly cited motivation is to show, by tracing back the state to a time before the agent was born, that the agent's actions are produced by factors that are clearly not under his control. I would prefer to put the point slightly differently. In my form of the construction we see that all human actions, in so far as they are physically characterizable, are deterministically explained by *exactly* the same factors — those that go to make up the state at $t = -10^{10}$ years — that explain the actions of sunflowers and everything else in the physical realm; and as for human mental life, either it is parasitic on the physical or else it makes no difference in producing the physical actions.

Libertarians are so-called because they want to go on to join to this construction further premises: if human actions are deterministically explained by exactly the same factors that explain the actions of sunflowers, then humans no more act freely than do sunflowers; or, alternatively, if human mental life is inefficacious in producing physical actions, or else is parasitic on the physical and the physical unfolds deterministically, then true 'choice' and 'decision' are illusions. While I feel a certain sympathy for these extra premises, I am not sure that I want to endorse either and align myself with the Libertarians, and I most certainly do not want to join the Woolite wing of the Libertarians who want to do a *modus tollens* on physical determinism, arguing that since determinism implies unfreedom and since we are free, determinism must be false. But to state what should now be obvious, it is not just the Libertarians who feel the crunch of determinism but anyone who wants to accord man a special place in nature on the grounds that, in contrast to inanimate objects and the lower life forms, we enjoy an autonomy in that what we do is up to us. This is why Schlick's tactic of trying to saddle his opponents with the indefensible 'determinism implies compulsion', already a questionable tactic when applied to Libertarians, is completely off and mark when applied to Autonomists.

There is an aspect of the tracing-back-the-state construction that has gone unremarked in the literature. Many of the fundamental physical laws we have studied are historically as well as futuristically determinstic. But we never see the construction presented in reverse form, tracing the state forward in time and then noting that by historical determinism the later state uniquely determines the present one. One obvious reason is the widespread acceptance of the notion that earlier states produce, cause, or bring about later states but not conversely. This notion was present in Laplace's original definition of Laplacian determinism and it crept into my presentation of the tracing-back construction. However, I now assert that despite all of the efforts of philosophers, the cause-effect relation in general and the notion of the direction of causation in particular remain so obscure that basing any conclusion on them makes the conclusion suspect. Suppose then that we reconsider the tracing-back construction, now taking care to understand futuristic determinism in a manner that does not entail any causal oomph. And suppose further that we agree with the claim of Ch. V that 'it is nomologically necessary that L' comes to no more than that L is part of the simplest and strongest system of occurrent regularities in the actual world. Then isn't much of the sting of determinism as it applies to human actions drawn? Perhaps, but drawing the sting in this way does nothing to soothe the Autonomist since even under the more scrupulous reading of determinism men and sunflowers are still in the same garden.

There is a related but perhaps more promising way to challenge the contention that men and sunflowers are in the same garden. Granted that if Laplacian determinism holds, the actions of men and sunflowers alike are deterministically explained by reference to the state at $t = -10^{10}$ years. But the actions of men are also explained by later states involving beliefs, preferences, motives, intentions, etc., and this constitutes a significant difference from sunflowers. The Libertarians and their allies the Autonomists seem to be tacitly appealing to the further principle that explanations using temporally prior states take precedence over or preempt explanations starting from later states, which principle may derive from the suspect notion that the earlier states causally produce the later ones. While I think that there is some justice to this charge, again I don't see that enough of the sting of determinism has been drawn to satisfy the Autonomist. To be concerned about autonomy one doesn't have to subscribe to the principle that the explanation starting from $t = -10^{10}$ years cancels or preempts explanations starting from later states; it is enough that the former exists.

It should be obvious by now that Popper's sense of 'indeterminism' does nothing to diminish the tension between determinism and freedom-autonomy, for Popper's 'indeterminism' is fully compatible with the deterministic scenario we have been imagining and it requires only that no embodied super-scientist can parallel the deterministic unfolding of events with precisely accurate forecasts (recall Ch. II).[1] It is only slightly less obvious that the real failures of ontological determinism I described for classical physics (Ch. III) and general relativistic physics (Ch. X) likewise provide cold comfort for the Libertarian. Whether any form of indeterminism will foster freedom and autonomy is the topic to which I now turn.

3. INDETERMINISTIC ACTIONS AND THE POSSIBILITY OF A SCIENCE OF HUMAN BEHAVIOR

Following the moves in the determinism-free will controversy is an exercise in frustration: whichever way we turn we are checked. We have been operating on the assumption of determinism. If we now turn away from this assumption then even worse results follow, or so we are told. B. F. Skinner argues that turning away from determinism is tantamount to abandoning the possibility of a science of human behavior:

If we are to use the methods of science in the field of human affairs, we must assume that behavior is lawful and determined. We must expect to discover that what a man does is the result of specifiable conditions and that once these conditions have been discovered, we can anticipate and to some extent determine his actions. (1953, p. 6)

An allied sentiment often found in Compatibilist tracts goes: if a man's actions are not determined but are merely spontaneous or random occurrences, then they cannot be said to be his actions as opposed to something that just happens to him.

The reader will immediately recognize that we are being presented with a false dichotomy: determinism vs. non-lawful behavior or determinism vs. spontaneity and randomness. In the preceding chapters I have tried to show just how important determinism, both as an empirical claim and as a guiding methodological principle, has been in the development of modern physics. But we have seen not the slightest reason to think that the science of physics would be impossible without determinism, and from the many examples studied we know that denying determinism does not push us over the edge of the lawful and into the abyss of the utterly chaotic and non-lawful.

Nevertheless there are valid points to the worry expressed in the passage from Skinner and in the little gloss following it. One of these points was made in a vivid, if somewhat overblown way, in R. E.

Hobart's (1966) famous article "Free Will as Involving Determination and Inconceivable Without It." If the rising of Alonzo's arm is to be counted as an action performed by Alonzo rather than a twitch or a jerk, then Alonzo's desire to raise his arm (or perhaps his willing his arm to go up) must in part determine the rising motion. To inject indeterminism here is not to inject freedom but to break the link that allows us to see Alonzo as the author of his action.

I want to suggest that one of the few valuable kernels in all of the chaff of the 'if and cans' literature is a reiteration of the same point in the language of causation. As an analysis of

(1)   Alonzo did A freely

we are offered

(2)   Alonzo could have done otherwise

And (2) in turn is analyzed as

(3)   Alonzo would have done otherwise if he had desired (willed) to do otherwise.

I suggest that (3) be viewed as a counterfactual test for the causal efficacy of Alonzo's desires (willings) in producing the action. Just such a test is used in the law of torts to establish causal responsibility, the formula being that but for the negligent action of the defendant the harm would not have resulted. Some philosophers want to promote the counterfactual test to the status of a full-blown analysis of causation (see Lewis (1973b)). The analysis founders, just as does every other attempt to explicate this murky notion, but this does not undermine the explanation of the role of (3) in discussions of free will. (Cause-effect terminology will keep popping up in the discussion of free will — the topic calls for it the way greasy food calls for catchup.)

As a test for freedom, however, (3) fails miserably, whether by the lights of the Compatibilist or the Libertarian. A Compatibilist of Schlick's persuasion would hold that Alonzo freely raised his arm if he was not acting under compulsion or coercion but simply following his natural desires. But this may be so while (3) is false. Suppose, for instance, that we can read Alonzo's desires from his brain waves and that if he shows the symptoms of not desiring to raise his arm we are prepared to step in and raise it for him with the help of an irresistibly strong mechanical arm.[2] We might try to save (3) as a necessary condition for freedom by claiming that if Alonzo had desired to reach down rather than up, then he would at least have put forth an effort to lower his arm and this is enough to satisfy the spirit of (3). But if you are willing to tolerate an even more science-fictiony scenario (and you must if you are going to talk free-will with philosophers), we can imagine that Alonzo's brain has been so wired by the Evil Scientist that upon receipt of the appropriate electrical impulse from the Evil One, Alonzo is unable to exert any muscular effort to move his arm. This may be overcome by substituting a mental effort for a muscular one, but that move runs the danger of turning (3) into a tautology. And it also brings us to the Libertarian's objection of (3) as a sufficient condition for freedom. If the hand of the Evil Scientist, or the invisible hand of determinism, controls Alonzo's desires and willings so that it was physically impossible in the circumstances for him to will or desire otherwise, then Alonzo was not free by the Libertarian's lights even though (3) may have been true. Relating this to Hobart, the Libertarian may agree that Alonzo's actions should be *conditionally* deterministic on his desires and willings but deny that, if free, his actions, desires, and willings are deterministic *simpliciter*. As so often happens with the determinism-free will problem, we have come full circle and are repeating ourselves.

After this excursion through one of the many byways of the free will problem, let us ask how indeterminism might help to resolve the concerns raised by the application of determinism to human actions. There are so many concerns and so many conflicting *desiderata* that this is a vague question, so for sake of definiteness I will impose three boundary conditions on a resolution of the determinism-free will problem.

(B1)   (Skinner) The resolution makes a science of human behavior possible.

(B2)   (Hobart) The resolution implies that the agent determines his behavior.

(B3)   (Autonomist) The resolution allows the agent to be autonomous in that his behavior is not a form of tropism.

We no sooner write down such conditions than we smell an impossibility. Indeterminism *per se* will not by itself secure the required autonomy. If, because of quantum mechanical considerations, the state at $t = -10^{10}$ years does not determine whether Alonzo will raise his arm in the next second but gives a probability of .734, then we have merely traded strict deterministic tropisms for probabilistic tropisms; men, sunflowers, radioactive decay, and photons impinging on half-silvered mirrors are all planted in the same probabilistic garden. A way to secure the desired autonomy is to somehow take the agent out of the flow of events, be it deterministic or indeterministic. Consistent with this interpretation of (B3) we may still be able to satisfy (B2), but not if we understand 'Alonzo determined his arm to go up' to mean that Alonzo was in such-and-such a physical-mental state and this state determined (in the sense we have been studying) via psycho-physical laws his subsequent action. What is required is a new kind of determinism in which the determiners are not states or events by agents or selfs. As C. D. Broad characterized the position,

the putting forth of a certain amount of effort in a certain direction at a certain time *is* completely determined, but is determined in a unique and peculiar way; it is literally determined by the *agent or self*, considered as a substance or continuant, and not by a total cause which contains as factors *events in* and *dispositions of* the agent. If this could be maintained, our puttings-forth of effort would be completely determined, but their causes would neither be events nor contain events as cause-factors. (1966, p. 157)

Broad goes on to try to show that agent determinism is a conceptual impossibility, but his 'proof' consists of no more than saying it just can't be (". . . in so far as an even *is* determined, an essential factor in its total cause must be other *events*").

What we can say with more confidence is that such a view is incompatible with Skinner's demand (B1). A recent proponent of agent causation, Roderick Chisholm, is almost forthright in drawing this moral:

This means that, in one very strict sense of the terms, there can be no science of man. If we think of science as a matter of finding out what laws happen to hold, and if the statement of a law tells us what kinds of events are caused by what other kinds of events, then there will be human actions which we cannot explain by subsuming them under any laws. We cannot say, 'It is causally necessary that, given such and such desires and beliefs, and being subject to such and such stimuli, the agent will do so and so'. For at times the agent, if he chooses, may rise above his desires and do something else instead. (1982, p. 33)

My only quarrel here is with Chisholm's qualifier "in one very strict sense of the terms." In any reasonable sense of the terms the conception of the self being offered is incompatible with a science of man. First, by construction there can be no natural laws at the level of agents or selfs.

Second, at the level of physical events natural laws become problematic as applied to human actions. Agent determinism is otiose if there is a perfect pre-established harmony between what the agent determines and the natural order of events, for then there would be no distinction between an event being determined by the agent vs. the event occurring naturally and undetermined by the agent. On the other hand, if the harmony is broken, one is at a loss to describe scientifically the difference the agent makes. One is at first tempted to say that the effect of the agent is a miracle — e.g., the agent determines atoms to swerve in contravention to the trajectories implied by natural laws. But in speaking thus we immediately contradict ourselves since what a law asserts must be true. Truth can be restored by building the exceptions into the law statement, but the modification removes the difference the agent was supposed to make and it saps both strength and simplicity, undermining the lawlikeness of the statement while supporting its truth.[3] Nor do I see that it helps to posit indeterministic probabilistic laws. Either the agent affects the relative frequencies of events or not, and either way we get a rerun of the same difficulties.[4] Agent determinism, so-called, pushes us toward the worst form of indeterminism — the absence of natural laws, deterministic or indeterministic, for human behavior. The self as something that is separable from and can rise above all momentary beliefs and desires and all longer term dispositions has been condemned in positivistic philosophy as a piece of metaphysical nonsense. I do not say that it is nonsense, but I do say that it cannot be the object of scientific scrutiny.

We have not ruled out the possibility that there is some other way to simultaneously satisfy the demands (B1)–(B3), and given the vagueness of the conditions no final proof will ever be forthcoming. But we gain strong evidence for the impossibility by watching two of the ablest philosophers of recent decades struggle valiantly but unsuccessfully to find a workable alternative. I do not want to spoil the pleasure the reader will find at perusing the attempts of Robert Nozick (1981) and Paul Meehl (1984) to provide a positive account of free will in a setting of physical indeterminism, but it may be useful to say a few words about how these attempts fit into the framework set out above.

Nozick, as I read him, is attempting to rehabilitate agent determinism or causation by bringing it down to the level of events. An agent's actions, on Nozick's account, are caused by his desires; so the spirit of Hobart's (B2) is satisfied without the need of a transcendent self as the cause of actions. Free will and autonomy are to be secured by maintaining that although the agent's actions are caused they are not causally determined since in exactly the same circumstances another action could have been performed. I think I detect in Nozick's description of this non-deterministic causation a ghost of the old agent determinism. The agent is supposed to somehow bestow weights on desires and so (indeterministically?) determine which desire prevails as the cause of the action; the old transcendent self which rises above the desires seems to have been replaced by a self which annoints desires. Meehl is concerned, as is Nozick, to rebut the charge that undetermined choices and actions are merely random or spontaneous occurrences. I agree with the rebuttal, but I get from neither Meehl nor Nozick any clear sense of how indeterminism makes the choice 'up to us' and secures for us an autonomy not enjoyed by indeterministic sunflowers while leaving us as objects of scientific enquiry. My suspicion that there is no such sense grows.

Some aspects of transcendent selfs which rise above desires and annointing selfs which bestow weights on desires can be made respectable by modeling them in terms of second and higher order desires (see Frankfurt (1982)). A second order desire may be a desire that our first order desires be such-and-so or that they have such-and-such strengths; or it may be a desire to act on one first order desire rather than another (called a volition in Harry Frankfurt's terminology). While the recognition of the hierarchical structure of desires has contributed valuable insights, I doubt that it holds a key to free will, for no matter how high up the hierarchy we climb the same concerns about autonomy follow us (see Slote (1980)). Moreover, when I reflect on decisions I have made under conditions of moral conflict — just the sort of circumstances where higher order desires come vividly into play — I do not get any special sense of having acted freely. On the contrary, upon looking back, there is a sense of inevitability, whether or not I gave into my first order appetite or overcame it by reflecting on how I ought to align my preferences to accord with the good or the just. By contrast, it seems on looking back that I was most free in those cases where no moral conflict was present and no higher order desire came into play (at least not consciously); e.g., I 'just chose' vanilla over chocolate ice cream. This inverts the usual doctrine that paradigm cases of free action are to be found in cases of conflict and angst. Thinking that 'rising above' first order desires by going to 2nd or nth level desires is the kind of rising above that confers freedom and autonomy is a mistake if my introspective evidence is any guide. Of course, introspected differences in degrees of freedom may not correspond to deep or interesting differences in psychology or physiology; still, it would be nice to have a more detailed and precise phenomenology of 'free choice' than is currently available.

4. CONCLUSION

Determinism leaves an indelible mark on every subject it brushes against. The mind-body problem is no exception. My characterization of physical determinism employs a conception of the physical that is so broad that it seems to swallow up the mental. In the terminology of events, causes, and effects that the mind-body theorists favor, a physical event in my sense is any event that fits into the spatio-temporal net and is thus a candidate for a cause or effect. Only the most farout dualists would deny that mental events are physical in this minimal sense. Of course, we still have a mind-body problem since it remains to be seen how those physical events we call mental are related to physical events in the narrower sense (e.g., those studied in standard physics). We have seen that physical determinism in the narrower sense strongly constrains any possible answer: mental events must be parasitic on physical events in the narrower sense, else differences in the former are not matched by any differences in the latter and are therefore inefficacious in bringing about physiological acts. Further pressure for an identity relation comes from the notion that parasitism without identity condemns mental events to the limbo of the epiphenominal, denying them a real causal role in producing the physiological act. I will leave it to those who think they know what a 'real causal role' is to adjudicate these issues. All I have to add is that both identity theorists and their opponents tend to operate with a false sense of how easy it is to establish identities in physics. All right thinking physicists believe that macro-thermodynamical quantities, states, and events are parasitic on the microscopic. But it turns out to be hard to characterize the parasitism in terms of identities (e.g., 'temperature is mean kinetic energy' is a glib over-simplification of a very complicated relation which may not in the end turn out to be a relation of identity). I am not suggesting that there are no special difficulties about the mental but only that a more careful look at problems of reduction in physics may point the way for more fruitful attacks on the mind-body question.

The more precisely science locates man in nature the more difficult it becomes to sustain a sense of autonomy for human actions.[5] As autonomy shrinks so does our sense of uniqueness and worth as well as the basis for a moral perspective on human action. As I have tried to indicate, this difficulty would arise even if the ultimate laws of nature proved to be non-deterministic, but since determinism poses the difficulty in its sharpest form it is appropriate to continue to speak of the determinism-free will problem. As a practical 'solution' I recommend the ostrich tactic: don't think too closely or too long on the issues raised here, and in daily life continue with the presumption that the 'I' that chooses and the self to which we attach value judgments are autonomous. Let those who want to call themselves philosophers bear the risk to their mental health that comes from thinking too much about free will.

NOTES

[1] Clerk Maxwell thought that instability made an opening for free will: "All of the great results produced by human endeavour depend on taking advantage of these singular [i.e., unstable] states when they occur" (Maxwell (1873), p. 443). But *how* do we take advantage? The miracle needed to contravene natural laws at an unstable point may be 'smaller' than that needed to contravene them at a stable point but it is nonetheless a miracle.

[2] See Frankfurt (1969) for criticisms of the 'could have done otherwise' formula as a necessary condition for freedom and responsibility.

[3] Recall the Mill-Ramsey-Lewis account of laws endorsed in Ch. V.

[4] But might not the agent help to produce the actually observed relative frequencies by 'reducing the superposition' as the Projection Postulate of quantum mechanics (recall Ch. XI) seems to require?

[5] Nozick (1981) and T. Nagel (1982) make this point in especially forceful ways. What is the explanation of the recent and growing trend among hard-headed analytic philosophers to reject the compatibilist solution?

SELECTED READINGS FOR CHAPTER XII

Clifford Williams' (1980) *Free Will and Determinism: A Dialogue* gives a quick 58 page introduction to some of the basic moves in the determinism-free will controversy. Bernard Berofsky's (1966) collection *Free Will and Determinism* contains many of the classic papers. Gary Watson's (1982) *Free Will* collects some of the more recent articles. The more intrepid readers will want to consult the chapter on "Free Will" from Robert Nozick's (1981) *Philosophical Explanations* and Paul Meehl's (1984) "Psychological Determinism or Chance: Configural Cerebral Autoselection as a Tertium Quid."

FINAL EXAM

(*Self-Administered*)

Instructions: Close the book and take a few deep breaths. Answer all
questions as best you can. Give yourself extra credit for
identifying the sources of the unattributed quotations, all of
whom are Famous Figures.

1. Why have some philosophers thought that the doctrine of determinism is necessarily true? Begin by citing some popular definitions of determinism.

2. Determinism means "ideally complete and precise predictability, given the momentary conditions, the pertinent laws, and the required mathematical techniques."

    Comment on this definition of determinism, bringing out its merits
and its demerits.

3. Evaluate the following:

    . . . the law of causality is neither right nor wrong, it can be neither proved nor
generally disproved. It is rather a heuristic principle, a sign-post (and to my mind
the most valuable sign-post we possess) to guide us in the motley confusion of
events and to show us the direction in which scientific research must advance in
order to attain fruitful results.

4. Karl Popper has claimed that "most systems of physics, including
classical physics and quantum physics, are indeterministic in perhaps an even more fundamental sense than the one usually
ascribed to the indeterminism of quantum physics . . ." Explain the
basis of this claim, and explain why Popper's sense of 'indeterminism' is but cold comfort to the Libertarian.

5. A leading philosopher of science has said that "few people would
wish to dispute the contention that classical mechanics is a deterministic theory whereas modern quantum mechanics is indeterministic." Without getting too deeply enmeshed in technical details,
explain why one ought to dispute the first part of this contention,
even if one rejects Popper's definition of determinism.

251

6. Show that in the case of $N = 3$, the Newtonian equations of motion for N gravitating point mass particles are deterministic, barring collisions.

7. Why is the Gödel incompleteness theorem not discussed in this book?

8. ... the fatalist asserts a causal discontinuity between present actions and the future world, where the non-fatalistic determinist asserts causal continuity here as everywhere else.... For the fatalist, no human causes can modify the future in any way, and resignation is the only rational course to follow. For the determinist, human efforts count as effectual causes along with all the other kinds of causes.

Write an essay on the relation between determinism and fatalism. Evaluate the above quotation in the light of the relation you have discerned.

9. There will be a sea battle tomorrow. True ☐; False ☐; Determined ☐; Undetermined ☐; Fated ☐; Not fated ☐; Why?

10. Answer *one* of the following:

(a) The Idealists have criticized the regularity account of laws on the grounds that

if the regularity view be right, all generalizations are nothing but sheer coincidences. If there is a connection between cause C and effect E so that one really explains or entails and does not just in fact precede the other, then the coincidence is indeed removed because there is now available a reason why E should always follow C, but not otherwise. On the regularity view there still just remains the brute fact that E always follows C, and that by itself is just as improbable as if an unweighted penny showed heads every time it was tossed ... The extraordinary unlikeliness of the generalization cannot be removed unless we suppose a logical connection between C and E. If a 'law' stands for nothing but the mere fact that E follows C, to posit such a general law because it is fulfilled in many cases is only to increase the improbability ... If C and E be logically connected so that the one entails the other the coincidence disappears, but otherwise how can it?

What is the 'logical connection' the Idealists are after? How can we come to know when and where it obtains? How might the regularity theorist reply to the above criticism?

(b) Evaluate the following analysis of laws of nature:

... for someone to treat a statement of the form 'if anything has Φ it has Ψ' as expressing a law of nature, it is sufficient (i) that subject to a willingness to explain

away exceptions he believes that in a non-trivial sense everything which in fact has
Φ has Ψ (ii) that his belief that something which has Φ has Ψ is not liable to
be weakened by the discovery that the object in question also has some other
property X, provided that (a) that X does not logically entail not-Ψ (b) that X is
not a manifestation of not-Ψ (c) that the discovery that something has X would
not in itself seriously weaken his belief that it had Ψ (d) that he does not regard
the statement 'if anything has Φ and not-X it has Ψ' as a more exact statement of
the generalization he was intending to express.

(c)

Generally speaking . . . a true scientific hypothesis will be regarded as a law of
nature if it has an explanatory function with regard to lower-level hypotheses or its
instances; vice versa, to the extent that a scientific hypothesis provides an explanation, to that extent will there be an inclination to endow it with the honourable
status of natural law.

In this claim true? If true, does it provide a means to a satisfying
analysis of 'law of nature', or does the notion of explanation itself
stand in need of an analysis which in turn will appeal to the concept
of law?

11. There is a maxim which is often quoted, that, 'The same cause will always produce
the same effects.' . . . What is really meant is that if the causes differ only as regards
the absolute time or the absolute place at which the event occurs, so likewise will
the effects.

Is this maxim essential to the physical sciences in general and to
determinism in particular? Explain. Why is it important to distinguish the above maxim from a second maxim asserting that 'Like
causes produce like effects'?

12. Answer *one* of the following:

(a) It has been said that Alan Turing set himself the task of

trying to specify exactly the most general possible notion of what a 'machine' is. In
fact, the definition he arrived at, now called a 'Turing machine,' was a central part
of his contribution to the theory of computing. Although fundamentally all a
Turing machine can do is jump from one discrete state to another by means of
very simple transition rules, Turing was able to show that such machines could do
anything one could reasonably expect of any machine or any human following
well-defined rules.

In what sense is this claim right? In what sense is it wrong?

(b) Alan Turing proposed that the digital computer could be used as a model for the human brain, and that a sufficiently powerful digital computer could so closely mimic the behavior of the brain that an interrogator could not tell from answers to questions put to X whether X is a digital machine or a man. Turing conceded that "the nervous system is certainly not a discrete machine . . . It may be argued that, this being so, one cannot expect to be able to mimic the behavior of the nervous system with a discrete state system." Here is his reply:

> It is true that a discrete state machine must be different from a continuous machine. But if we adhere to the conditions of the imitation game, the interrogator will not be able to take advantage of this difference. The situation can be made clearer if we consider some other simpler continuous machine. A differential analyzer will do very well . . . It would not be possible for a digital computer to predict exactly what answers the differential analyzer would give to a problem, but it would be capable of giving the right sort of answer. For instance, if asked to give the value of π (actually about 3.1416) it would be reasonable to choose at random between the values 3.12, 3.13, 3.14, 3.15, 3.16 with the probabilities of 0.05, 0.15, 0.55, 0.19, 0.06 (say). Under these circumstances it would be very difficult for the interrogator to distinguish the differential analyzer from the digital computer.

Does Turing's reply really show that the digital computer is an adequate model for simple *deterministic* continuous state machines, much less for brains?

13. To what extent can a deterministic device generate 'random numbers'? Distinguish the cases of finite vs. infinite random sequences, and discuss separately discrete vs. continuous state systems.

14. Define some of the ergodic properties of classical dynamical systems and discuss how they are relevant to explaining how 'chance' can emerge from a deterministic evolution. Give some examples of classical systems and sketch reasons for thinking that these systems do (or do not) exhibit the ergodic properties.

15. . . . the replacement of Newtonian mechanics by Einstein's special (1905) and general (1916) theories of relativity did nothing to upset the deterministic character of physics. Newton's laws of mechanics turned out to be not quite correct, so they had to be replaced by some revised laws of mechanics, but ones that were no less deterministic.

This claim is: True □; False □; True but misleading □; False but informative □. Why?

16. Provide an interesting but not obviously false formulation of the 'cosmic censorship hypothesis' for general relativistic cosmological models. Either demonstrate that the hypothesis is true, or else give an explicit counterexample. Detail the implications of your answer for determinism.

17. The essentially non-deterministic character of quantum mechanics rests on the principle of indeterminacy, sometimes called the uncertainly relation, first stated in 1927 by Werner Heisenberg. It says, roughly, that, for certain pairs of magnitudes called 'conjugate' magnitudes, it is impossible in principle to measure both at the same instant with high precision ... [So with momentum p and position q it] asserts that, if we try to measure p precisely, that is, make Δp very small, we cannot at the same instant measure q precisely, that is, make Δq very small.

Is this interpretation of the uncertainty principle correct? Supposing for sake of argument that it is, what does it imply about indeterminism?

18. Derive some form of the Bell inequalities and exhibit a quantum state in which the quantum probabilities are in violation of the inequality. For what views in the foundations of quantum mechanics does this violation toll? Is a deterministic interpretation of the theory among them?

19. In the epilogue to ___ the famous Russian novelist ___ wrote:

Every man, savage and sage alike, however incontestably reason and experience may prove to him that it is impossible to imagine two different courses of action under precisely the same circumstances, yet feels that without this meaningless conception (which constitutes the essence of freedom) he cannot conceive of life. He feels that, however impossible it may be, it is so; seeing that without that conception of freedom, he would not only be unable to understand life, but could not live for a single instant. ... To history the recognition of freewills of men as forces able to influence historical events, that is, not subject to laws, is the same as would be to astronomy the recognition of freewill in the movements of heavenly bodies ... If there is a single human action due to freewill, no historical law exists, and no conception of historical events can be formed.

Does the laboring of this theme detract from the artistic merit of the novel? Would the author have been so confident of the theme if the novel had been set in the US rather than Russia?

20. The principle of free will says: '*I* produce my volitions.' Determinism says: 'My volitions are produced by *me*.' Determinism is free will expressed in the passive voice.

The freedom of the will consists in the impossibility of knowing actions that still lie in the future.

Refute these pieces of sophistry.

BIBLIOGRAPHY

Aberth, O.: (1980), *Computable Analysis* (New York: McGraw Hill).

Alexander, H. G. (ed.): (1956), *The Leibniz-Clarke Correspondence* (Manchester: Manchester University Press).

Anderson, C. M. and Von Baeyer, H. C.: (1972), "Almost Circular Orbits in Classical Action-at-a-Distance Electrodynamics," *Physical Review D5*, 802—813.

Anderson, J. L.: (1967), *Principles of Relativity Physics* (New York: Academic Press).

Aristotle: (1908), *Metaphysica*, Vol. 8 of J. A. Smith and W. D. Ross (eds.), *The Works of Aristotle* (Oxford: Oxford University Press).

Armstrong, D. M.: (1978), *Universals and Scientific Realism*, 2 Vols. (Cambridge: Cambridge University Press).

Armstrong, D. M.: (1983), *What is a Law of Nature?* (Cambridge: Cambridge University Press).

Arnold, V. I. and Avez, A.: (1968), *Ergodic Problems of Classical Mechanics* (New York: W. A. Benjamin).

Aspect, A., Dalibard, J., and Roger, D.: (1982), "Experimental Test of Bell's Inequalities Using Variable Analyzers," *Physical Review Letters 49*, 1804—1807.

Ayer, A. J.: (1956), "What is a Law of Nature?" *Revue Internationale de Philosophie 10*, 144—165.

Beem, J. K.: (1980), "Minkowski Space-Time Is Locally Extendible," *Communications in Mathematical Physics 72*, 273—275.

Belasco, E. P. and Ohanian, H. C.: (1969), "Initial Conditions in General Relativity: Lapse and Shift Formulation," *Journal of Mathematical Physics 10*, 1503—1507.

Belinfante, F. J.: (1973) *A Survey of Hidden Variable Theories* (New York: Pergamon Press).

Bell, J. S.: (1964), "On the Einstein Podolsky Rosen Paradox," *Physics 1*, 195—200.

Bell, J. S.: (1966), "On the Problem of Hidden Variables in Quantum Mechanics," *Reviews of Modern Physics 38*, 447—475.

Berofsky, B. (ed.): (1966), *Free Will and Determinism* (New York: Harper and Row).

Berofsky, B.: (1968), "The Regularity Theory," *Noûs 2*, 315—340.

Born, M.: (1971), *The Born-Einstein Letters* (London: Macmillan).

Born, M. and Jordan, P.: (1925), "Zur Quantenmechanik," *Zeitschrift für Physik 34*, 858—888.

Bowie, G. L.: (1972), "An Argument Against Church's Thesis," *Journal of Philosophy 70*, 66—76.

Braithwaite, R. B.: (1960), *Scientific Explanation* (New York: Harper Torch Books).

Broad, C. D.: (1966), "Determinism, Indeterminism, and Libertarianism," in B. Berofsky (1966).

Brush, S.: (1976), *The Kind of Motion We Call Heat*, Vol. II (Amsterdam: North-Holland).

Budic, R. and Sachs, R. K.: (1976), "Deterministic Space-Times," *General Relativity and Gravitation 7*, 21—29.

Cahn, S. M.: (1967), *Fate, Logic, and Time* (New Haven, CT: Yale University Press).

Carnap, R.: (1964), "The Methodological Character of Theoretical Concepts," in H. Feigl (ed.), *Minnesota Studies in the Philosophy of Science*, Vol. I (Minneapolis, MN: University of Minnesota Press).

Carslaw, H. S.: (1921), *Introduction to the Mathematical Theory of Heat Conduction in Solids* (London: Macmillan).

Cartwright, N.: (1977), "The Sum Rule Has Not Been Tested," *Philosophy of Science 44*, 107—113.

Chatin, G.: (1975), "Randomness and Mathematical Proof," *Scientific American 232* (May), 47—52.

Chisholm, R.: (1982), "Human Freedom and the Self," in G. Watson (1982).

Christodoulou, D. and Francaviglia, M.: (1979), "The geometry of the thin-sandwich problem," in J. Ehlers (ed.) (1979).

Church, A.: (1940), "On the Concept of a Random Sequence," *Bulletin of the American Mathematical Society 46*, 130—135.

Clarke, C. J. S.: (1973), "Local Extensions in Singular Space-Times," *Communications in Mathematical Physics 32*, 205—214.

Clarke, C. J. S.: (1976), "Space-Time Singularities," *Communications in Mathematical Physics 49*, 17—23.

Clauser, J. F., Horne, M. A., Shimony, A., and Holt, R. A.: (1969), "Proposed Experiment to Test Local Hidden Variable Theories," *Physical Review Letters 23*, 880—883.

Clauser, J. F. and Horne, M. A.: (1974), "Experimental consequences of objective local theories," *Physical Review D10*, 526—535.

Clauser, J. F. and Shimony, A.: (1978), "Bell's Theorem: experimental tests and implications," *Reports on Progress in Physics 41*, 1881—1927.

Cohen, L.: (1966a), "Generalized Phase-Space Distribution Functions," *Journal of Mathematical Physics 7*, 781—786.

Cohen, L.: (1966b), "Can Quantum Mechanics be Formulated as a Classical Probability Theory?" *Philosophy of Science 33*, 317—322.

Courant, R. and Friedrichs, K. O.: (1976), *Supersonic Flow and Shock Waves* (New York: Springer-Verlag).

Currie, D. G.: (1966), "Poincaré Invariant Equations of Motion for Classical Particles," *Physical Review 142*, 817—824.

Cutland, N.: (1980), *Computability* (Cambridge: Cambridge University Press).

Daneri, A., Loinger, A. and Prosperi, G. M.: (1962), "Quantum Theory of Measurement and Ergodicity Conditions," *Nuclear Physics 33*, 297—319.

DeWitt, B. S. and Graham, N. (eds.): (1973), *The Many-Worlds Interpretation of Quantum Mechanics* (Princeton, NJ: Princeton University Press).

D'Espagnat, B.: (1976), *Conceptual Foundations of Quantum Mechanics*, 2nd Ed. (Reading, MA: W. A. Benjamin).

Dirac, P. A. M.: (1926), "Heisenberg's Quantum Mechanics and the Hydrogen Atom," *Proceedings of the Royal Society (London) A110*, 561—579.

Doob, J. L.: (1941), "Probability as Measure," *Annals of Mathematical Statistics 12*, 206—214.

Dretske, F. I.: (1977), "Laws of Nature," *Philosophy of Science 44*, 248—268.

Driver, R.: (1963), "A Two-Body Problem of Classical Electrodynamics: The One Dimensional Case," *Annals of Physics 21*, 122—142.

Driver, R.: (1969), "A Backwards Two-Body Problem of Classical Relativistic Electrodynamics," *Physical Review 178*, 2051—2057.

Driver, R.: (1977), *Ordinary and Delay Differential Equations* (New York: Springer-Verlag).

Driver, R.: (1978), "Can the future influence the present?" *Physical Review D19*, 1098—1107.

Dunn, J. M.: (1973), "A Truth Value Semantics for Modal Logic," in H. Leblanc (ed.), *Truth, Syntax, and Modality* (New York: North-Holland).

Earman, J.: (1977), "Till the End of Time," in J. Earman *et al.* (eds.), (1977).

Earman, J.: (1979), "Was Leibniz a Relationist?" in P. French et al. (eds.), *Midwest Studies in Philosophy*, Vol. IV (Minneapolis, MN: University of Minnesota Press).

Earman, J.: (1985), "Locality, Non-Locality, and Action-at-a-Distance: A Skeptical Review of Some Philosophical Dogmas," to appear in P. Achinstein and R. Kargon (eds.), *Theoretical Physics in the 100 Years since Kelvin's Baltimore Lectures* (Cambridge, MA: M.I.T. Press).

Earman, J.: (1986), "Meaningfulness and Invariance," to appear in P. Ehrlich and C. W. Savage (eds.), *Minnesota Studies in the Philosophy of Science* (Minneapolis, MN: University of Minnesota Press).

Earman, J. (ed.): (1983), *Minnesota Studies in the Philosophy of Science*, Vol. X (Minneapolis, MN: University of Minnesota Press).

Earman, J., Glymour, C., and Stachel, J. (eds.): (1977), *Minnesota Studies in the Philosophy of Science*, Vol. VIII (Minneapolis, MN: University of Minnesota Press).

Earman, J., and Glymour, C.: (1982), "What Is Classical Electromagnetism?" pre-print.

Earman, J., Glymour, C. and Rynasiewicz, R. A.: (1982), "On Writing the History of Special Relativity," in P. Asquith and T. Nickles (eds.), *PSA 1982* (East Lansing, MI: Philosophy of Science Association).

Earman, J. and Norton, J.: (1986), "What Price Space-Time Substantivalism?" pre-print.

Earman, J. and Shimony, A.: (1968), "A Note on Measurement," *Nuovo Cimento 54B*, 332—334.

Eberhard, P. H.: (1977), "Bell's Theorem without Hidden Variables," *Nuovo Cimento 38B*, 75—79.

Eells, E.: (1983), "Objective Probability Theory Theory," *Synthese 57*, 387—442.

Ehlers, J. (ed.): (1979), *Isolated Gravitating Systems in General Relativity* (New York: North-Holland).

Einstein, A.: (1905), "Zur Elektrodynamik bewegter Körper," *Annalen der Physick 17*, 891—921.

Einstein, A.: (1961), *Relativity, The Special and General Theory* (New York: Bonanza Press).

Einstein, A., Podolsky, B., and Rosen, N.: (1935), "Can Quantum-Mechanical Description of Reality be Considered Complete?" *Physical Review 47*, 777—780.

Elliott, J. P. and Dawber, P. G.: (1979), *Symmetry in Physics*, 2 vols. (Oxford University Press).

Ellis, G. F. R., and King, A. R.: (1974), "Was the Big Bang a Whimper?" *Communications in Mathematical Physics 38*, 119—156.

Ellis, G. F. R., and Schmidt, B. G.: (1977), "Singular Space-Times," *General Relativity and Gravitation 8*, 915—953.

Everett, H.: (1957), "'Relative State' Formulation of Quantum Mechanics," *Reviews of Modern Physics 29*, 454—462.

Ewing, A. C.: (1974), *Idealism, A Critical Survey* (London: Methuen and Co.).

Fattorini, H. D.: (1983), *The Cauchy Problem* (Reading, MA: Addison-Wesley).

Feigl, H. and Brodbeck, M. (eds.): (1953), *Readings in the Philosophy of Science* (New York: Appleton-Century-Crofts).

Feigl, H.: (1953), "Notes on Causality," in H. Feigl and M. Brodbeck (1953).

Feller, W.: (1968), *An Introduction to Probability Theory and Its Applications*, Vol. I (New York: John Wiley).

Fetzer, J. H.: (1981), *Scientific Knowledge* (Dordrecht: D. Reidel).

Fetzer, J. H., and Nute, D. E.: (1979), "Syntax, Semantics and Ontology: A Probabilistic Causal Calculus," *Synthese 40*, 453—495.

Fetzer, J. H., and Nute, D. E.: (1980), "A Probabilistic Causal Calculus: Conflicting Conceptions," *Synthese 44*, 241—246.

Fine, A.: (1970), "Insolubility of the Quantum Measurement Problem," *Physical Review D2*, 2783—2787.

Fine, A.: (1971), "Probability in Quantum Mechanics and Other Statistical Theories," in M. Bunge (ed.), *Problems in the Foundations of Physics* (New York: Springer-Verlag).

Fine, A.: (1973), "Probability and the Interpretation of Quantum Mechanics," *British Journal for the Philosophy of Science 24*, 1—37.

Fine, A.: (1977), "Conservation, the Sum Rule and Confirmation," *Philosophy of Science 44*, 95—106.

Fine, A.: (1981), "Correlations and Physical Locality," P. Asquith and R. N. Giere (eds.), *PSA 1980* (East Lansing, MI: Philosophy of Science Association).

Fine, A.: (1982a), "Hidden Variables, Joint Probability, and the Bell Theorems," *Physical Review Letters 48*, 291—295.

Fine, A.: (1982b), "Joint Distributions, Quantum Mechanics, and Commuting Observables," *Journal of Mathematical Physics 23*, 1306—1310.

Fine, A.: (1982c), "Some Local Models for Correlation Experiments," *Synthese 50*, 279—294.

Fine, A. and Teller, P.: (1978), "Algebraic Constraints on Hidden Variables," *Foundations of Physics 8*, 629—636.

Fine, T. L.: (1973), *Theories of Probability* (New York: Academic Press).

van Fraassen, B. C.: (1977), "Relative Frequencies," *Synthese 34*, 133—166.

van Fraassen, B. C.: (1980), *The Scientific Image* (Oxford: Oxford Universtiy Press).

van Fraassen, B. C.: (1981), "Essences and Laws of Nature," in R. Healey (ed.) *Reduction, Time and Reality* (Cambridge: Cambridge University Press).

Frankfurt, H. G.: (1969), "Alternative Possibilities and Moral Responsibility," *Journal of Philosophy 66*, 829—839.

Frankfurt, H. G.: (1982), "Freedom of the Will and the Concept of a Person," in G. Watson (1982).

Friedman, M.: (1983), *Foundations of Space-Time Theories* (Princeton, NJ: Princeton University Press).

Garrido, L. (ed.): (1983), *Dynamical Systems and Chaos* (New York: Springer-Verlag).

Geroch, R.: (1968), "What Is a Singularity in General Relativity?" *Annals of Physics 48*, 526—540.

Geroch, R.: (1977), "Prediction in General Relativity," in J. Earman (ed.) (1977).

Geroch, R.: (1984), "The Everett Interpretation," *Noûs 18*, 617—633.

Geroch, R. and Horowitz, G. T.: (1979), "Global Structure of Space-Times," in S. W. Hawking and W. Israel (eds.) (1979).

Gerver, J. L.: (1984), "A Possible Model for a Singularity without Collisions in the Five Body Problem," *Journal of Differential Equations 52*, 76—90.

Ghirardi, G. and Rimini, A.: (1982), "Some Topics in the Quantum Theory of Measurement," in A. Avez (ed.), *Dynamical Systems and Microphysics* (New York: Academic Press).

Giere, R.: (1973), "Objective Single-Case Probabilities and the Foundations of Statistics," in P. Suppes *et al.* (eds.), *Logic, Methodology, and Philosophy of Science IV* (Amsterdam: North-Holland).

Giere, R.: (1976), "A Laplacian Formal Semantics for Single Case Propensities," *Journal of Philosophical Logic 5*, 321—353.

Gleason, A. M.: (1957), "Measures on the Closed Subspaces of a Hilbert Space," *Journal of Mathematics and Mechanics 6*, 885—893.

Glymour, C.: (1971), "Determinism, Ignorance, and Quantum Mechanics," *Journal of Philosophy 68*, 744—751.

Glymour, C.: (1977), "The Sum Rule is Well Confirmed," *Philosophy of Science 44*, 86—94.

Glymour, C.: (1980), *Theory and Evidence* (Princeton, NJ: Princeton University Press).

Gödel, K.: (1949), "An Example of a New Type of Cosmological Solution to Einstein's Field Equations," *Reviews of Modern Physics 21*, 447—450.

Goodman, N.: (1955), *Fact, Fiction and Forecast* (Cambridge, MA: Harvard University Press).

Grzegorczyk, A.: (1955), "Computable functionals," *Fundamenta Mathematica 42*, 168—202.

Grzegorczyk, A.: (1957), "On the definitions of computable real continuous functions," *Fundamenta Mathematica 44*, 61—71.

Gudder, S.: (1970), "On Hidden-Variable Theories," *Journal of Mathematical Physics 11*, 431—436.

Hadamard, J.: (1952), *Lectures on Cauchy's Problem in Linear Partial Differential Equations* (New York: Dover Publications).

Hartman, P. and Wintner, A.: (1950), "On the Solutions of the Equation of Heat Conduction," *American Journal of Mathematics 72*, 367—395.

Hawking, S. W.: (1979), "Comments on Cosmic Censorship," *General Relativity and Gravitation 10*, 1047—1049.

Hawking, S. W. and Ellis, G. F. R.: (1973), *The Large Scale Structure of Space-Time* (Cambridge: Cambridge University Press).

Hawking, S. W. and Penrose, R.: (1970), "Singularities of Gravitational Collapse and Cosmology," *Proceeding of the Royal Society (London) A 314*, 529—548.

Hawking, S. W. and Israel, W. (eds.): (1979), *General Relativity* (Cambridge: Cambridge University Press).

Healey, R. A.: (1984), "How Many Worlds?" *Noûs 18*, 591—616.

Held, A. (ed.): (1980), *General Relativity and Gravitation*, 2 vols. (New York: Plenum Press).

Helleman, R. H. G.: (1980), "Self-generating chaotic behavior in non-linear mechanics," in E. C. D. Cohen (ed.), *Fundamental Problems in Statistical Mechanics V* (New York: North-Holland).

Hellman, G.: (1982), "Einstein and Bell: Strengthening the Case for Microphysical Randomness," *Synthese 53*, 445—460.

Hill, R. N.: (1967), "Instantaneous Action-at-a-Distance in Classical Relativistic Mechanics," *Journal of Mathematical Physics 8*, 201—220.

Hill, R. N.: (1967a), "Canonical Formulation of Relativistic Mechanics," *Journal of Mathematical Physics 8*, 1756—1773.

Hill, R. N.: (1970), "Instantaneous Interaction Relativistic Dynamics for Two Particles in One Dimension," *Journal of Mathematical Physics 11*, 1918—1937.

Hill, R. N.: (1982), "The Origins of Predictive Relativistic Mechanics," in J. Llosa (ed.) (1982).

Hirschman, I. I.: (1952), "A Note on the Heat Equation," *Duke Mathematical Journal 19*, 487—492.

Hobart, R. E.: (1966), "Free Will as Involving Determination and Inconceivable Without It," in B. Berofsky (ed.) (1966).

Hochberg, H.: (1981), "Natural Necessity and Laws of Nature," *Philosophy of Science 48*, 386—399.

Hoering, B.: (1969), "Indeterminism in Classical Physics," *British Journal for the Philosophy of Science 20*, 247—255.

Hsing, D.-P. K.: (1977), "Existence and uniqueness theorem for the one-dimensional backwards two-body problem of electrodynamics," *Physical Review D16*, 974—982.

Hume, D.: (1973), *Treatise of Human Nature* (Oxford: Oxford University Press).

Hume, D. (1975), *Enquiry Concerning Human Nature* (Oxford: Clarendon Press).

Jackson, A. S.: (1960), *Analogue Computation* (New York: McGraw Hill).

James, W.: (1956), "The Dilemma of Determinism," in *The Will to Believe* (New York: Dover Publications).

Jammer, M.: (1974), *The Philosophy of Quantum Mechanics* (New York: Wiley-Interscience).

Jauch, J. M.: (1968), *Foundations of Quantum Mechanics* (Reading, MA: Addison-Wesley).

John, F.: (1982), *Partial Differential Equations* (New York: Springer-Verlag).

Jones, R.: (1981), "Is General Relativity Generally Relativistic?" in P. D. Asquith and R. N. Giere (eds.), *PSA 1980*, Vol. 2 (East Lansing, MI: Philosophy of Science Association).

Kaempffer, F. A.: (1965), *Concepts in Quantum Mechanics* (New York: Academic Press).

Kerner, E. H. (ed.): (1972), *The Theory of Action-at-Distance in Relativistic Particle Dynamics* (New York: Gordon and Breach).

Kerner, E. H. and Sutcliffe, W. G.: (1970), "Unique Hamiltonian Operators via Feynman Path Integrals," *Journal of Mathematical Physics 11*, 391—393.

Kneale, W.: (1949), *Probability and Induction* (Oxford: Oxford University Press).

Kochen, S. and Specker, E. P.: (1967), "The Problem of Hidden Variables in Quantum Mechanics," *Journal of Mathematics and Mechanics 17*, 59—87.

Kundt, W.: (1967), "Non-Existence of Trousers Worlds," *Communications in Mathematical Physics 4*, 143—144.

Kuryshkin, V. V.: (1977), "Uncertainty Principle and the Problems of Joint Coordinate-Momentum Probability Density in Quantum Mechanics," in W. C. Price and S. S. Chissik (eds.), *The Uncertainty Principle and the Foundations of Quantum Mechanics* (New York: John Wiley).

Kyburg, H.: (1974), "Propensities and Probabilities," *British Journal for the Philosophy of Science 25*, 358—375.

Landau, L. D. and Lifshitz, E. M.: (1959), *Fluid Mechanics* (Oxford: Oxford University Press).

Lanford, O. E.: (1975), "Time Evolution of Large Classical Systems," in J. Moser (ed.), (1975).

Lanford, O. E.: (1977), "An Introduction to the Lorenz System," in D. Ruelle (ed.) (1977).

Laplace, P. S.: (1820), *Théorie analytique des probabilités* (Paris: V. Courcier).

Laplace, P. S.: (1951), *A Philosophical Essay on Probabilities* (New York: Dover Publications).

Lax, P. D.: (1973), *Hyperbolic Systems of Conservation Laws and the Mathematical Theory of Shock Waves* (Philadelphia, PA: Society for Industrial and Applied Mathematics).

Lebowitz, J. L. and Penrose, O.: (1973), "Modern Ergodic Theory," *Physics Today 23* (Feburary), 23—29.

Leray, J.: (1934), "Essai sur la movement d'un liquide vis quenx emplissant l'espace," *Acta Mathematica 63*, 193—248.

Lewis, D.: (1973a), *Counterfactuals* (Cambridge, MA: Harvard University Press).

Lewis, D.: (1973b), "Causation," *Journal of Philosophy 70*, 556—567.

Lichtenberg, A. J. and Lieberman, M. A.: (1983), *Regular and Stochastic Motion* (New York: Springer-Verlag).

Llosa, J. (ed.): (1982), *Relativistic Action at a Distance: Classical and Quantum Aspects* (New York: Springer-Verlag).

London, F. W. and Bauer, E.: (1939), *La théorie de l'observation en méchanique quantique* (Paris: Hermann et Cie).

Lorenz, E. N.: (1963), "Deterministic Nonperiodic Flow," *Journal of Atmospheric Science 20*, 130—141.

Malament, D.: (1984), "Newtonian Gravity, Limits, and the Geometry of Space," to appear in R. Colodny (ed.), *Quarks and Quasars* (Pittsburgh, PA: University of Pittsburgh Press).

Malament, D. and Zabell, S.: (1980), "Why Gibbs Phase Space Averages Work — The Role of Ergodic Theory," *Philosophy of Science 47*, 339—349.

Markus, L. and Meyer, K.: (1974), *Generic Hamiltonian Dynamical Systems Are Neither Integrable Nor Ergodic* (Providence, RI: American Mathematical Society).

Martin-Löf, P.: (1966), "The definition of a random sequence," *Information and Control 9*, 62—619.

Martin-Löf, P.: (1969), "The Literature on von Mises' Kollektivs Revisited," *Theoria 35*, 12—37.

Martin-Löf, P.: (1970), "On the Notion of Randomness," in A. Kino (ed.), *Intuitionism and Proof Theory* (Amsterdam: North-Holland).

Mather, J. N. and McGehee, R.: (1975), "Solutions of the Collinear Four-Body Problem Which Become Unbounded in a Finite Time," in J. Moser (ed.), (1975).

Maxwell, J. C.: (1873), "Does the progress of Physical Science tend to give any advantage to the opinion of Necessity (or Determinism) over that of the Contingency of Events and the Freedom of the Will?", quoted in L. Campbell and W. Garnett: (1882), *The Life of James Clerk Maxwell* (London: Macmillan).

Maxwell, J. C.: (1890), "On Boltzmann's Theorem on the average distribution of energy in a system of material points," in *The Scientific Papers of James Clark Maxwell*, Vol. 2 (Cambridge: Cambridge University Press).

Maxwell, J. C.: (1920), *Matter and Motion* (New York: Dover Publications).

Mazur, S.: (1963), "Computable Analysis," *Dissertations Mathematicae 33*, 4—110.

McLenaghan, R. G.: (1969), "An explicit determination of the empty space-times on which the wave equation satisfies Huygens' principle," *Proceedings of the Cambridge Philosophical Society 65*, 139—155.

Meehl, P.: (1984), "Psychophysical Determinism or Chance: Configural Cerebral Autoselection as a Tertium Quid," to appear in *Minnesota Studies in the Philosophy of Science*, Vol. XI (Minneapolis, MN: University of Minnesota Press).

Mellor, D. H.: (1971), *The Matter of Chance* (Cambridge: Cambridge University Press).

Mellor, D. H.: (1974), "In Defence of Dispositions," *Philosophical Review 83*, 153—181.

Mill, J. S.: (1904), *A System of Logic* (New York: Harper and Row).

Miller, A. I.: (1981) *Albert Einstein's Special Theory of Relativity* (Reading, MA: Addison-Wesley).

Minkowski, H.: (1909), "Raum und Zeit," *Physikalishe Zeitschrift 20*, 104—111. English translation reprinted in Perrett and Jeffrey (1923).

von Mises, R.: (1941), "On the Foundations of Probability and Statistics," *Annals of Mathematical Statistics 12*, 191—205.

von Mises, R.: (1957), *Probability, Statistics, and Truth* (New York: Dover Publications).

von Mises, R.: (1964), *Mathematical Theory of Probability and Statistics* (New York: Academic Press).

Møller, C.: (1972), *The Theory of Relativity* 2d. ed. (Oxford: Oxford University Press).

Moncrief, V.: (1982), "Neighborhoods of Cauchy Horizons in Cosmological Space-Times with One Killing Field," *Annals of Physics 141*, 83—102.

Montague, R.: (1962), "Towards a General Theory of Computability," in *Logic and Language: Studies Dedicated to Professor Rudolf Carnap* (Dordrecht: D. Reidel).

Montague, R.: (1974), "Deterministic Theories," in R. H. Thomason (ed.), *Formal Philosophy* (New Haven, CT: Yale University Press).

Moser, J. (ed.): (1975), *Dynamical Systems, Theory and Applications* (New York: Springer-Verlag).

Müller zum Hagen, H. and Seifert, H.-J.: (1979), "The Characteristic Initial Value Problem in General Relativity," in J. Ehlers (ed.) (1979).

Nagel, E.: (1953), "The Causal Character of Modern Physical Theory," in H. Feigl and M. Brodbeck (eds.) (1953).

Nagel, E. (1961), *The Structure of Science* (New York: Harcourt, Brace, and World).

Nagel, T.: (1982), "Moral Luck," in G. Watson (1982).

von Neumann, J.: (1955), *Mathematical Foundations of Quantum Mechanics* (Princeton, NJ: Princeton University Press).

Nordtvedt, K.: (1974), "Causality and Metrical Properties of Matter in a Two-Metric Field Theory of Gravity," in W. B. Rolnick (ed.), *Causality in Physical Theories* (New York: American Institute of Physics).

Norton, J.: (1984), "How Einstein found his field equations," *Studies in the History and Philosophy of Science 14*, 255—276.

Nozick, R.: (1981), *Philosophical Explanations* (Cambridge, MA: Belknap Press).

Oxtoby, J. C. and Ulam, S. M.: (1941), "Measure Preserving Homeomorphisms and Metrical Transitivity," *Annals of Mathematics 42*, 847—920.

Payne, L. E.: (1975), *Improperly Posed Problems in Partial Differential Equations* (Philadelphia, PA: Society for Industrial and Applied Mathematics).

Pauli, W.: (1958), *Theory of Relativity* (New York: Dover Publications).

Pearle, P.: (1976), "Reduction of the state vector by a non-linear Schrödinger equation," *Physical Review D13*, 857—864.

Penrose, R.: (1965), "A Remarkable Property of Plane Waves in General Relativity," *Reviews of Modern Physics 37*, 215—220.

Penrose, R.: (1967), *An Analysis of the Structure of Space-Time*, mimeo (Princeton University).

Penrose, R.: (1978), "Singularities of Space-Time," in N. Lebovitz, W. Reid and P. Vandervoot (eds.), *Theoretical Principles in Astrophysics and Relativity* (Chicago, IL: University of Chicago Press).

Perrett, W. and Jeffrey, G. B.: (1923), *The Principle of Relativity* (New York: Dover Publications).

Péter, R.: (1957), "Rekursivität und Konstruktivität," in A. Heyting (ed.), *Constructivity in Mathematics* (Amsterdam: North-Holland).

Poincaré, H.: (1906), "Sur la dynamique de l'electron," *Rendiconti del Circolo Matematica di Palerno 21*, 129—175.

Pollard, H.: (1966) *Mathematical Introduction to Celestial Mechanics* (Englewood-Cliffs, NJ: Prentice-Hall).

Popper, K.: (1959), "The Propensity Interpretation of Probability," *British Journal for the Philosophy of Science 10*, 25—42.

Popper, K. R.: (1962), "The propensity interpretation of probability, and the quantum theory," in S. Körner (ed.) *Observation and Interpretation in the Philosophy of Physics* (New York: Dover Publications).

Popper, K. R.: (1971), "Of Clouds and Clocks," in *Objective Knowledge* (Oxford: Oxford University Press).

Popper, K. R.: (1982), *The Open Universe* (Totowa, NJ: Rowman and Littlefield).

Pour-El, M. B.: (1974), "Abstract Computability and Its Relation to the General Purpose Analogue Computer," *Transactions of the American Mathematical Society 199*, 1—28.

Pour-El, M. B. and Caldwell, J.: (1975), "On a simple definition of computable function of a real variable," *Zeitschrift für Mathematische Logik und Grundlagen der Mathematik 21*, 1—19.

Pour-El, M. B. and Richards, I.: (1979), "A computable ordinary differential equation which possesses no computable solutions," *Annals of Mathematical Logic 17*, 61—90.

Pour-El, M. B. and Richards, I.: (1981), "The Wave Equation With Computable Initial Data Such That Its Unique Solution is Not Computable," *Advances in Mathematics 39*, 215—239.

Pour-El, M. B. and Richards, I.: (1983), "Noncomputability in Analysis and Physics: A Complete Determination of the Class of Noncomputable Linear Operators," *Advances in Mathematics 48*, 44—74.

Putnam, H.: (1970), "Is Logic Empirical?" in R. S. Cohen and M. W. Wartofsky (eds.), *Boston Studies in the Philosophy of Science*, Vol. 5 (Dordrecht: D. Reidel).

Putnam, H.: (1976), "How to Think Quantum Logically," in P. Suppes (eds.), *Logic and*

Probability in Quantum Mechanics (Dordrecht: D. Reidel).

Quine, W. V. O.: (1976), *The Ways of Paradox* (Cambridge, MA: Harvard University Press).

Ramsey, F. P.: (1978), *Foundations of Mathematics* (Atlantic Highlands, NJ: Humanities Press).

Redhead, M.: (1980), "Experimental Tests of the Sum Rule," *Philosophy of Science 48*, 50—64.

Reichenbach, H.: (1954), *Nomological Statements and Admissible Operations* (Amsterdam: North-Holland).

Reichenbach, H.: (1971), *The Theory of Probability* (Berkeley, CA: University of California Press).

Rescher, N.: (1970), *Scientific Explanation* (New York: Free Press).

Rivier, D. C.: (1951), "On a One-to-One Correspondence between Infinitesimal Canonical Transformations and Infinitesimal Unitary Transformations," *Physical Review 83*, 862—863.

Rogers, H.: (1967), *Theory of Recursive Functions and Effective Computability* (New York: McGraw Hill).

Ruelle, D. (ed.): (1977), *1976 Duke Turbulence Conference* (Duke University Mathematics Series III).

Ruelle, D.: (1977a), "Statistical Mechanics and Dynamical Systems," in Ruelle (1977).

Ruelle, D.: (1980), "Strange Attractors," *Mathematical Intelligencer 2*, 126—137.

Ruelle, D.: (1981), "Differential Dynamical Systems and the Problem of Turbulence," *Bulletin of the American Mathematical Society 5*, 29—42.

Russell, B.: (1936), "Determinism and Physics," *Proceedings of the University of Durham Philosophical Society*.

Russell, B.: (1948), *Human Knowledge: Its Scope and Limitations* (Simon and Schuster).

Russell, B.: (1953), "On the Notion of Cause with Applications to the Free-Will Problem," in H. Feigl and M. Brodbeck, (1953).

Rynasiewicz, R. A.: (1981), *Varieties of Eliminability of Theoretical Terms and the Empirical Content of Theories*, Ph.D. dissertation, University of Minnesota.

Rynasiewicz, R. A.: (1983), "Falsifiability and Semantic Eliminability," *British Journal for the Philosophy of Science 34*, 225—241.

Sachs, R. K.: (1962), "On the Characteristic Initial Value Problem for General Relativity," *Journal of Mathematical Physics 3*, 908—911.

Salmon, W.: (1971), "Determinism and Indeterminism in Modern Science," in J. Feinberg (ed.), *Reason and Responsibility* (Encino, CA: Dickenson Publishing Co.).

Salvioli, A.: (1972), "On the Theory of Geometric Objects," *Journal of Differential Geometry 7*, 247—278.

Saari, D. G.: (1973), "Improbability of Collisions in Newtonian Gravitational Systems.II," *Transactions of the American Mathematical Society 182*, 351—368.

Saari, D. G.: (1977), "A Global Existence Theorem of the Four-Body Problem of Newtonian Mechanics," *Journal of Differential Equations 26*, 80—111.

Schlick, M.: (1966), "When Is a Man Responsible?", in Berofsky (ed.) (1966).

Schnorr, C.: (1971), "A Unified Approach to the Definition to Random Sequence," *Mathematical Systems Theory 5*, 246—258.

Schnorr, C.: (1971a), *Zufälligkeit und Wahrscheinlichkeit* (New York: Springer-Verlag).

Schouten, J. A.: (1954), *Ricci-Calculus* 2d. ed. (New York: Springer-Verlag).

Shannon, C.: (1941), "Mathematical Theory of the Differential Analyzer," *Journal of Mathematics and Physics 20*, 337—354.

Shepherdson, J. C. and Sturgis, H. E.: (1963), "Computability of Recursive Functions," *Journal of the ACM 10*, 217—235.

Shewell, J. R.: (1959), "On the Formation of Quantum-Mechanical Operators," *American Journal of Physics 27*, 16—21.

Shimony, A.: (1974), "Approximate Measurement in Quantum Mechanics, II," *Physical Review D9*, 2321—2323.

Shimony, A.: (1984), "Contextual Hidden Variable Theories and Bell's Inequalities," *British Journal for the Philosophy of Science 35*, 25—45.

Skinner, B. F.: (1953), *Science and Human Behavior* (New York: Free Press).

Sklar, L.: (1973), "Unfair to Frequencies," *Journal of Philosophy 67*, 255—266.

Skyrms, B.: (1980), *Causal Necessity* (New Haven, CT: Yale University Press).

Slote, M.: (1980), "Understanding Free Will," *Journal of Philosophy 77*, 136—151.

Smart, J. J. C.: (1968), *Between Philosophy and Science* (New York: Random House).

Sperling, H. J.: (1970), "On the real singularities of the *N*-body problem," *Journal für die Reine und Angewandte Mathematik 245*, 15—40.

Stairs, A.: (1982), "Quantum Logic and the Lüders Rule," *Philosophy of Science 49*, 422—436.

Stapp, H. P.: (1971), "*S*-Matrix Interpretation of Quantum Theory," *Physical Review D3*, 1303—1320.

Stein, H.: (1977), "Some Pre-History of General Relativity" in J. Earman (ed.) (1977).

Stein, H.: (1984), "The Everett Interpretation of Quantum Mechanics: Many Worlds or None?" *Noûs 18*, 636—652.

Streater, R. F.: (1975), "Outline of axiomatic relativistic quantum field theory," *Reports on Progress in Physics 38*, 771—846.

Suchting, W. A.: (1974), "Regularity and Law," in R. S. Cohen and M. W. Wartofsky (eds.) *Boston Studies in the Philosophy of Science*, Vol. 14 (Dordrecht: D. Reidel).

Swoyer, C.: (1982), "The Nature of Natural Laws," *Australasian Journal of Philosophy 60*, 203—223.

Synge, J. L.: (1964), *Relativity: The Special Theory* 2d, ed. (Amsterdam: North-Holland).

Szekeres, P.: (1975), "Quasispherical Gravitational Collapse," *Physical Review D12*, 2941—2948.

Taylor, R.: (1967), "Determinism," *The Encyclopedia of Philosophy*, Vol. II (New York: Macmillan and Free Press).

Taylor, R.: (1983), *Metaphysics* (Englewood-Cliffs, NJ: Prentice-Hall).

Teller, P.: (1979), "Quantum Mechanics and the Nature of Continuous Physical Magnitudes," *Journal of Philosophy 76*, 345—361.

Teller, P.: (1983), "The Projection Postulate as a Fortuitous Approximation," *Philosophy of Science 50*, 413—431.

Temam, R.: (1983), *Navier-Stokes Equations and Nonlinear Functional Analysis* (Philadelphia, PA: Society for Industrial and Applied Mathematics).

Thomas, W. J.: (1973), "Doubts About Some Standard Arguments for Church's Thesis," in R. Bogdan and I. Niiniluoto (eds.), *Logic, Language, and Probability* (Dordrecht: D. Reidel).

Tipler, F. J.: (1974), "Rotating Cylinders and the Possibility of Global Causality Violations," *Physical Review D9*, 2203—2206.

Tipler, F. J.: (1980), "General Relativity and Eternal Return," in F. J. Tipler (ed.) *Essays in General Relativity* (New York: Academic Press).

Tipler, F. J., and Clarke, C. J. S., and Ellis, G. F. R.: (1980), "Singularities and Horizons — A Review Article," in A. Held (ed.), (1980).

Tondl, L.: (1973), *Scientific Procedures* (Dordrecht: D. Reidel).

Tooley, M. A.: (1977), "The Nature of Laws," *Canadian Journal of Philosophy 7*, 667—698.

Tourlakis, G. J.: (1984), *Computability* (Reston, VA: Reston Publishing Co.).

Turing, A.: (1937), "On Computable Numbers with an Application to the Entscheidungsproblem," *Proceedings of the London Mathematical Society 42*, 230—265.

Ville, J.: (1939), *Étude critique de la notion collectif* (Paris: Gauthier-Villars).

Wald, A.: (1938), "Die Widerspruchsfreiheit des Kollektivbegriffes," *Actualities Scientifique et Industrielles 735*, 79—99.

Wald, R.: (1984), *General Relativity* (Chicago, IL: University of Chicago Press).

Watson, G. (ed.): (1982), *Free Will* (Oxford: Oxford University Press).

Weyl, H.: (1950), *The Theory of Groups and Quantum Mechanics* (New York: Dover Publications).

Wheeler, J. A.: (1957), "Assessment of Everett's 'Relative State' Formulation of Quantum Theory," *Reviews of Modern Physics 29*, 151—153.

Wheeler, J. A.: (1962), *Geometrodynamics* (New York: Academic Press).

Wheeler, J. A.: (1964), "Mach's Principle as boundary condition for Einstein's equations," in H.-Y. Chiu and W. F. Hoffmann (eds.), *Relativity and Gravitation* (New York: W. A. Benjamin).

Wheeler, J. A.: (1968), "Superspace and the Nature of Quantum Geometrodynamics," in C. M. DeWitt and J. A. Wheeler (eds.), (1978), *Battelle Recontres* (New York: W. A. Benjamin).

Wheeler, J. A.: (1977), "Singularity and Unanimity," *General Relativity and Gravitation 8*, 713—715.

Wheeler, J. A. and Zurek, W. H. (eds.): (1983), *Quantum Theory and Measurement* (Princeton, NJ: Princeton University Press).

Whitrow, G. J. and Morduch, G. E.: (1965), "Relativistic Theories of Gravitation," in A. Beer (ed.) *Vistas in Astronomy*, Vol. 6 (New York: Pergamon Press).

Widder, D. V.: (1975), *The Heat Equation* (New York: Academic Press).

Wigner, E. P.: (1952), "Die Messung quantenmechanischer Operatoren," *Zeitschrift für Physik 133*, 101—108.

Wigner, E. P.: (1961), "Remarks on the Mind-Body Question," in I. J. Good (ed.) *The Scientist Speculates* (London: Heinemann).

Wigner, E. P.: (1970), "On Hidden Variables and Quantum Mechanical Probabilities," *American Journal of Physics 39*, 1005—1009.

Williams, C.: (1980), *Free Will and Determinism: A Dialogue* (Indianapolis, IN: Hackett).

Wilson, H. V. R.: (1955), "Causal Discontinuity in Fatalism and Indeterminism," *Journal of Philosophy 52*, 70—72.

Wintner, A.: (1947), *The Analytical Foundations of Celestial Mechanics* (Princeton, NJ: Princeton University Press).

Wittgenstein, L.: (1961), *Tractatus Logico-Philosophicus* (London: Routledge and Kegan Paul).

Yanase, M.: (1961), "Optimal Measuring Apparatus," *Physical Review 123*, 666–668.

Yodzis, P., Seifert, H. J. and Müller zum Hagan, H.: (1974), "On the occurrence of naked singularities in general relativity," *Communications in Mathematical Physics 34*, 135–148.

INDEX OF NAMES